Beyond Sex Differences
Genes, Brains and Matrilineal Evolution

ERIC B. KEVERNE
University of Cambridge

CAMBRIDGE
UNIVERSITY PRESS

University Printing House, Cambridge CB2 8BS, United Kingdom

One Liberty Plaza, 20th Floor, New York, NY 10006, USA

477 Williamstown Road, Port Melbourne, VIC 3207, Australia

4843/24, 2nd Floor, Ansari Road, Daryaganj, Delhi – 110002, India

79 Anson Road, #06-04/06, Singapore 079906

Cambridge University Press is part of the University of Cambridge.

It furthers the University's mission by disseminating knowledge in the pursuit of education, learning, and research at the highest international levels of excellence.

www.cambridge.org
Information on this title: www.cambridge.org/9781108416856
DOI: 10.1017/9781108242080

© Cambridge University Press 2017

This publication is in copyright. Subject to statutory exception and to the provisions of relevant collective licensing agreements, no reproduction of any part may take place without the written permission of Cambridge University Press.

First published 2017

Printed in the United Kingdom by Clays, St Ives plc

A catalogue record for this publication is available from the British Library.

ISBN 978-1-108-41685-6 Hardback

Cambridge University Press has no responsibility for the persistence or accuracy of URLs for external or third-party internet websites referred to in this publication and does not guarantee that any content on such websites is, or will remain, accurate or appropriate.

Contents

	Preface	*page* vii
	Acknowledgements	xiii
1	On the Genetic Origin of Sex Differences	1
2	Epigenetics: The Gene Environment Interface	19
3	Genomic Imprinting: Matrilineal Regulatory Control Over Gene Expression	52
4	Puberty: Developmental Reorganisation of Sex Differences in Body and Mind	86
5	Mother–Infant Bonding: The Biological Foundations for Social Life and Cultures	111
6	Brain and Placenta: The Coming Together of Two Distinct Generations	145
	The Epigenetic Landscape in the Evolutionary Ascent of the Matriline: Concluding Overview	172
	Index	207

Preface
Understanding the Biology of Sex Differences: Mother Leads the Way

It is widely understood and well accepted that males and females differ biologically and in many ways. How our views of biological differences arose, and how these views have shaped our attitudes to the opposite sex has, however, been a long and ever-changing narrative in need of reconsideration. The ancient understandings of sex differences has, without doubt, nurtured some unrealistic male-biased opinions. Such male-biased evaluations of sex differences are nowhere more clearly illustrated than by the writings of men who have been considered eminent and respected in this field. In tracing this narrative historically, it appears that the early Athenian view of sex differences and human biology was that women were merely the incubators of the male seed. Greek philosophers advocated that 'mother is not the parent of that which is called the child, but it is mother that nurtures the seed of the child that grows. The parent is he that plants the seed' (Blundell, 1995). Aristotle, despite his infinite wisdom, claimed women merely supplied the material origins of humanity and males supplied the higher causes, thus making 'males superior to females' (Lange, 1983). It is by no means clear as to which, if any, biological knowledge of sex differences could have given rise to these opinions of women, other than the mothers' commitment to producing and caring for children. Of course, we rarely, if ever, hear the opinions or views of ancestral women on this subject, probably because women were seldom considered to be worthy writers or philosophers, and were primarily recognised by society through and for the man they married. Thucydides proclaimed that 'women's greatest joy is to be talked about among men, whether this be in praise or in blame'. Moving forward in time to the thirteenth century, but not progressing in ideology, St Thomas Aquinas, an Italian theologian, philosopher and the pre-eminent spokesman for Catholic reason and divine revelation, postulated that the production of women was a defect in the production of men.

In more recent times, the renowned psychiatrist, Sigmund Freud, theorised that femininity was an immature stage of masculine development. Freud, although a medical doctor, was primarily focused on the psyche in his work, but even those medics who worked on the physiology of reproduction were also inclined to consider female reproduction as the 'default' state. Alfred Jost, a receiver of many scientific awards for his work on the biology of sexual differentiation, and who subsequently became Secretary to the French Academy of Sciences 1991–2001, adopted this same viewpoint. He carried out animal experiments, which showed that male rats castrated before puberty developed a female phenotype (Josso, 2008); hence the conclusion that females represented a default state arising from males. This, to some extent, matched the human clinical findings where a biological insensitivity to male hormones also produces a 'female phenotype'. Even as recently as 1975, the concluding paragraph on sex determination in a medical text on reproduction stated that 'maleness means mastery of the Y chromosome over the X, the medulla over the cortex, androgen over oestrogen. So physiologically speaking, there is no justification for believing in the equality of the sexes; vive la différence'.

In the British Victorian era, sex and reproduction was a 'no-go area' for biological research. Biological research was primarily an all-male preserve with select women attending discussion meetings, but more as spectators than contributors. The Victorian convention was that science knowledge in women was not feminine, and certainly to be discouraged by the male scientists. Even the great libertarian, Thomas Huxley, noted that a minority of women, suitably educated, might become fit 'companions' of men, but not their 'competitors'. Their proper role was to be more concerned with the scientist than with his science (Richards, 1989). In 1864, Huxley claimed to have found anatomical evidence for female inferiority. In his Hunterian lectures, he described the structural differences in the brains of men and women, noting that 'the cerebral convolutions are simpler in women than men'. He went on to conclude that 'in every excellent character, whether mental or physical, the average woman is inferior to the average man in the sense of having that character less in quantity and lower in quality'. The great Charles Darwin followed Huxley's lead by arguing on evolutionary grounds 'that the higher education of women could have no long term impact on their social evolution and was, strictly speaking, a waste of resources'.

Paul Broca (1824–1880), an eminent anthropologist of his times, took a slightly different stance, stating that 'Although nature has not made men and women equal, these facts do not afford grounds for refusing to educate women or give them the same civil and political rights. In the name of Justice, law and custom should not add to the biological burdens that weigh women down. Women's subordination to man was natural and eternal and any attempt to revolutionise the education and status of women on the assumption of an imaginary sexual equality would induce a perturbation in their evolution' (Broca, 1868).

I have been somewhat surprised and dismayed to find these misogynous writings by such eminent philosophers and renowned scientists from the past, who were recognised for their rational thinking and wisdom. I would have liked to believe that such views were influenced and biased by opinions that were embedded in the culture of their times. However, many of these eminent male citizens helped to formulate this culture during their lifetime. Theirs was a weak and less-than-objective attempt to come to terms with what represents our current understanding of the biology and psychology of male/female differences today. My own view, after decades of studying the brain and behaviour, is that there are indeed biological sex differences but, importantly, these sex differences in humans are not purely driven by the male sex hormone testosterone, nor are they simply down to the genes on the Y chromosome. On the contrary, it is primarily the mother's genome and epigenome (the matrilineal genome) that has played the leading role in determining mammalian and, in turn, our own successful evolutionary progression. This matrilineal progression has finally resulted in the complex organisation of today's human societies. In understanding how this lead has been established by the matriline, I have focussed on genetics, epigenetics, brain development and behaviour. Each of these areas of study have further benefitted from their consideration within an evolutionary and genomic framework. In mammals, this has required taking account of the disproportionate biological role undertaken by the matriline in order to achieve reproductive success. Of course, males have benefitted equally, and share the advantages accruing from such progressive changes. However, for the most part, the patrilineal role has been mainly that of an essential but compliant passenger along our evolutionary road to biological success. Clearly, there is a mismatch between the historic social 'cognitive' view of the female and today's knowledge of the biological reality of gender differences.

What are the studies of the last 40 years that have challenged such biological delusions of male dominance and leadership, and provided a more informed and balanced view on the biology of male versus female sex differences? The two most important and informative areas of biological research to clarify our modern way of thinking have been in genetics and in neuroscience. My aim in this monograph is to draw upon this recent information in order to provide a more balanced overview of the role, indeed the leading role, which the matriline has contributed to our biological success, and how this has had impact on the evolutionary advancement of the human brain. I have spent my research career investigating the brain, its gross and functional anatomy, and the molecular genetics that regulate its development. It is clear to me that a comprehensive understanding of brain function requires that we take into account how the brain relates to the body, and equally important, how this relationship differs between males and females. Getting brain function into perspective further requires an understanding of its evolution. Evolutionary studies have provided a perspective that is relevant to all aspects of development and function of the brain, from genetics to physiology and anatomy, and provides a logical insight as to how brain structure and function have changed over the millennia.

The success of mammalian reproduction owes a great deal to maternal in-utero development (viviparity), as indeed does evolution of the brain (see Chapter 4). In the context of in-utero viviparity, the female brain has developed mechanisms for regulation of post-partum care and mother–infant bonding. These same mechanisms have evolved to underpin social relationships and social living, as exemplified by the female bonded social groups of the large-brained monkeys and apes (Curley & Keverne, 2005). The extended period of care and commitment provided by social living has also provided a lengthy period for postnatal brain development and growth. Moreover, such brain development takes place within an environment dominated by complex social relationships. Social living thus heralded a massive evolutionary increase in those parts of the brain that take on social and executive functions. Importantly, human reproduction has witnessed a level of emancipation for maternal care and sexual behaviour from the determining action of the body's biological messengers, the hormones. So important has been the extended postnatal period of growth in the evolution of the human executive brain that the hormonal biology of reproduction has been succeeded by growth of the forebrain in guaranteeing reproductive success. From this pivotal stage forwards, the brain has been developing the capacity to shape the direction of its own evolutionary destiny. Only humans can override the basic instincts of hunger, sex and aggression. Only humans can undertake lifetime vows of chastity, or become committed mothers without undertaking pregnancy, as indeed can men, although the male social predisposition in this direction is rarely so devotional.

My aim in writing this book is to draw upon recent studies that inform and provide a balanced overview of the role, indeed the leading role, which the matrilineal genome has contributed to the evolutionary development of differences between males and females. The impact that the matriline has made on mammalian evolution is particularly reflected in the genetics and epigenetics of brain and placental development. The success of Motherhood, as exemplified by mammals, has been driven by in-utero placentation. This in turn has provided the coexistence of two genomes, those of mother and foetus, which have provided a platform for intergenerational genomic co-adaptation. It is motherhood, not sex, which has provided the selection pressures for achieving mammalian reproductive success. The mother–infant relationship has underpinned the evolutionary development of large brains, while the neural mechanisms for bonding and attachment have, in turn, provided the biological foundations for underpinning complex social organisations.

REFERENCES

Blundell, S. (1995). *Women in Ancient Greece*. First (US) edn. Cambridge, MA: Harvard University Press.

Broca, P. (1868). On anthology. *Anthropol. Rev.* 6: 35–52.

Curley, J. P. & Keverne, E. B. (2005). Genes, brains and mammalian social bonds. *Trends Ecol. Evol.* 20: 561–67.

Josso, N. (2008). Professor Alfred Jost: the builder of modern sex-differentiation. *Sex. Dev.* 1: 55–63.

Lange, L. (1983). Woman is not a rational animal: on Aristotle's biology of reproduction. In: S. Harding & M. B. Hintikka (eds.), *Discovering Reality*. Dordrecht: D. Reidel Publishing Company, pp. 1–15.

Richards, E. J. (1989). Huxley and woman's place in science: the 'woman question' and the control of Victorian anthropology. In: J. R. Moore (ed.), *History, Humanity and Evolution: Essays for John C. Greene*. New York, NY: Cambridge University Press.

Acknowledgements

Diane Pearce, my former secretary, has been of immeasurable help and patience during the preparation of this book. I could not have managed without her help. My thanks go out to all those students and post-doctoral research appointees who have worked with me over the decades, and have contributed to many of the studies reported here. They have played an essential role in the formulation of many of the ideas presented in these chapters.

Especial thanks go to my friends and colleagues Professor Joe Herbert and Professor Azim Surani, who provided the intellectual excitement and stimulated many of my studies, and especially my ways of thinking about the brain and development. My daughter Jessica also gets a special 'thank you' for the feedback, encouragement and support that stirred me out of my retirement lethargy.

1   On the Genetic Origin of Sex Differences

Our children inherit a set of 23 chromosome pairs carrying the genetic information from mother and from father. Out of these 46 human chromosomes, the majority are matched for maternal and paternal genes, the so-called homologous pairs of genes. The exception is seen for the sex chromosome pair. In sons, the male Y-chromosome combines with the female X-chromosome, while daughters have two copies of the X-chromosome, one from the mother and one from the father. The genes on one of the two X-chromosomes that daughters inherit are silenced to ensure that X-gene dosage for sons and daughters always match.

The sex chromosomes took their name from early cytogenetic studies when it was observed that males have one chromosome, named the Y-chromosome, which is very much smaller than its partner, the X-chromosome. Male mammals possess an XY sex chromosome complement (heterogametic sex) and females have two XX chromosomes (homogametic sex). These cytogenetic studies revealed that the possession of a Y-chromosome determined the development of the male's testes, and in the absence of a Y-chromosome in the female, an ovary would develop. From this early stage of development forwards, sexual differentiation was considered to be dependent on the hormones produced by the testes or the ovaries. Of course, these steroid hormones have never been considered to be directly coded for by specific genes, but they are an epigenetic product of gene activity, and in turn they are able to activate genes by way of their protein receptors. These steroid receptors are coded for by the genome, and these receptor genes reside on many chromosomes within many different cell types. It is important to appreciate that these early cytogenetic studies represented a description of what takes place, and not an explanation of the mechanisms involved. There were many gaps in our knowledge, not least of which concerned those genes residing on the sex chromosomes which coded for producing the female and male gonads. The gonads in turn ensured production of the sex hormones. Investigating the evolutionary history of sex determination across different species was not a great deal of help for our understanding of the origins of mammalian genetic sex determination. In amphibia, reptiles and birds, the female is the heterogametic sex. Moreover, in some reptiles, there is a lack of any chromosomal differences between the sexes, and ambient temperature during the period of egg incubation has been found to be the epigenetic sex-determining factor (Matsumoto et al., 2013). There are also certain exceptions to normal gonadal development in mammals, with sex reversal in wild rodent species that lack any Y-chromosome (Jiminez et al., 2013). Evidence for what the missing genes from the Y-chromosome might be have come from studies of a rodent found in Japan (the Amami Spiny rat). This Spiny rat has no Y-chromosome (XO/XO), nor the sex-determining *SRY* gene (Kuroiwa et al., 2011). In the Spiny rat, mechanisms for sex determination are underpinned by multiple copies of the autosomal *Cbx2* gene in the male, but not in females. In the male Spiny rat, these multiple copies of the *Cbx2* gene are responsible for producing the testes.

A single gene on the mammalian Y-chromosome, the so-called *SRY* male sex-determining gene, was identified in 1990. This *SRY* gene is now known to be primarily responsible for testes development, and thereby production of androgenic hormones which promote the creation of a male phenotype and one which differs from that of the female (Cortez et al., 2014). This presence of a single Y-chromosome has not always provided the underpinning for sex differences even among different mammalian species. In the evolutionary history of our recent mammalian ancestors, represented in modern times by the Australian Duck-Billed Platypus, sex determination is a function of many genes across multiple chromosomes

(Ferguson-Smith & Rens, 2010). So what have been the advantages to providing the modern male mammal with a unique Y-chromosome and a unique *SRY* sex-determining gene? Other than male sex determination, few advantages have accrued from possession of a single Y-chromosome. Indeed, the Y-chromosome has suffered extensive gene loss, with only 3% of its ancestral genes functionally surviving. Moreover, it has recently been found that with the use of assisted reproduction in mice, live offspring can be produced lacking the entire Y-chromosome long arm (Yamauchi *et al.*, 2014). Moreover, these progeny developed as males by using only two genes from the Y-chromosome, namely *SRY* and the spermatogenic proliferation factor *Eif2s3y*.

Two main processes are believed to have resulted in the evolutionary loss of genes from the Y-chromosome. Genes in that region of the Y-chromosome, which is non-recombining with the X-chromosome, are inactivated and lost as a consequence of point mutations, DNA deletions and the accumulation of nucleotide insertions. Moreover, the higher rate of mutations on the Y-chromosome is due to the many rounds of replication undertaken during sperm production, and the absence of DNA repair enzymes. Some experts have even postulated that the human Y-sex chromosome may eventually disappear (Sun & Heitman, 2012). There is, however, no need for male concern in the near future. This 3 per cent of remaining Y-chromosome genes are functionally coherent, enriched for maintaining gene dosage stability, and have remained relatively stable for the past 25 million years.

Gene repression as well as gene expression in mammalian gonadal development has become a complex and important process in the understanding of sex differences. Recent studies have found that so-called polycomb repressive complexes play a crucial role in regulating repression of certain genes involved in sexual development (Katoh-Fukui *et al.*, 2012). Interestingly, male development results from the repression of female ovarian-determining genes followed by the expression of the male *SRY* gene. Female ovarian development is very much an active process, while male testes development depends upon repressing female ovarian development. It is the expression of *Cbx2* gene in the male which represses maternal transcription from taking place and thereby development of the ovary. Mapping analysis of this gene's transcription has identified some 1600 targets for *Cbx2*, many of which code for proteins known to be involved in disorders of sexual development (Eid *et al.*, 2015). Hence, genetic determination of sex differences is now known to be extremely complex, and not just dependent on the *SRY* gene in males (Figure 1.1). *SRY* is essential for male testes development, while the *Cbx2* complex allows the testes-determining *Sox9* gene to be expressed by the inhibition of *FoxL2*, another gene actively engaged in female ovarian development. *Cbx2* is expressed in a parent of origin manner, with multiple copies being produced in the male, but not in the female (Tardat *et al.*, 2015).

The mammalian X-chromosome, in contrast to the Y-chromosome, has been proposed to enable important selection pressures to operate for mammalian evolution due to the accumulation

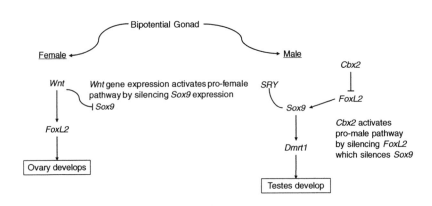

See ref. Biason-Lauber and Chabrissier, 2015

FIGURE 1.1 Overview of ovaries and testes development. The bipotential gonad develops into the testes by the *Cbx2* gene blocking the *FoxL2* gene, thereby allowing the *SRY* gene to activate *Sox9* expression and testes development.

of placental and brain-expressed genes. The X-chromosome has also acquired and amplified gene families that are expressed in the male testes. Moreover, the X-chromosome of males is always inherited from their mother, and has thereby gained a specialisation for male reproduction via the matriline and by the acquisition of new genes. These new genes acquired on the X-chromosome have tended to be of importance for regulation of gene networks that govern the expression of gene targets across multiple chromosomes throughout the genome. Thus, the sum total of the male's genetic capacity to determine masculinity is no longer solely present on the single Y-chromosome, although the Y-chromosome still plays an integral role in this process through the male *SRY* gene and the spermatogonadial proliferation factor.

The extensive gene loss from the male Y-chromosome has taken place in multiple steps. This loss was never sufficient to prevent the Y-chromosome from combining with the X-chromosome, but it certainly reduced the ability for natural selection to operate on this Y-chromosome. The Y-chromosome is always inherited through father to son, and is thus unique in that this chromosome never passes through the female germline. The consequences of such inheritance provide a focus for selection pressures to be exclusively specialised for, and primarily beneficial to, male functions. However, the majority of mutational changes which do occur to genes on the Y-chromosome are not advantageous. There is a sound biological explanation for this finding; because of the multiple cell divisions the sperm germ cells undertake, then the greater is the risk for mutational errors arising (Bachtrog, 2013). Males are particularly vulnerable to genetic errors because billions of sperm are produced by the multiple cell divisions that occur in the testes during the reproductive lifetime of a male. When compared with the relatively few cell divisions that are required to produce the full complement of female eggs (oocytes) prior to female birth, then there is considerably greater risk for accumulating male germline mutations. Not only are there more of these mutations occurring in the male germline, but there is no opportunity for DNA repair in the male germline. Unlike the X-chromosome, the male Y-chromosome does not possess a homologous partner. Thus, there is no duplicate for the Y-chromosome against which mismatch DNA repair can be undertaken. This only remains possible for those very few duplicate genes that are present on both the X- and Y-chromosomes, but these genes are also under regulatory control by the maternal X-chromosome and are advantageous to both sexes.

It is clear that the X-chromosome has retained or acquired many of the former male sex-related genes, including the receptor for testosterone, while other former Y-chromosome genes have been taken over by the remaining 22 autosomal chromosomes. The male hormone testosterone, the most potent of the male androgenic hormones, determines male secondary sexual characteristics by acting on the so-called androgen receptor. The gene for this receptor is expressed from the female X-chromosome in males and is, therefore, always inherited via mother (Migeon et al., 1981). During mammalian evolution, the demasculinisation of the Y-chromosome became further associated with feminisation of the X-chromosome, through X-linked genes focussing expression to the ovaries (Bion & Toniolo, 2000). Thus, unlike the degenerate Y-chromosome, evolutionary positive selection has accompanied the early evolutionary progression for genes on the X-chromosome, which males always inherit from the female.

Although the Y-chromosome is essential to developing a male phenotype through the development of the testes and their production of testosterone, this hormone itself functions through the androgen receptor. As mentioned, the gene coding for this receptor has become incorporated into the female X-chromosome (Migeon et al., 1981). Thus, a clinical syndrome of sexual development is seen to occur in XY males that become feminised due to a default in the androgen receptor gene on their X-chromosome. This dysfunction is known as 'testicular feminisation syndrome' (shortened to Tfm) (Wang et al., 2014). Androgens are still produced by the testes in Tfm males, but none of the somatic body tissues, or brain neurons, have the receptor that enables them to respond to this testosterone. Hence, the male brain and body of Tfm males fail to respond to male testosterone, and develops to respond for a feminine phenotype. These Tfm subjects display a feminine gender identity and, if they are reared as girls, they date and marry men. We can conclude from these findings that there are no genes on the male Y-chromosome that are sufficient in their own right to enable the development of male secondary sexual characteristics, or to induce psychological masculinity. Indeed, the mapping of brain responses to sexually arousing imagery that is seen in Tfm males provides the same responses as those for the brain of women.

It is difficult to understand why or to rationalise the claim of early reproductive biologists for 'females' to represent a default state. Their choice of terminology was inappropriate, and with a greater knowledge of the genetics underpinning sexual differentiation, this view has actually been proved to be incorrect. Indeed, it would be more accurate to define masculinity as dependent on the female germline. Certainly, the *SRY* male sex-determining gene on the Y-chromosome is essential for male reproduction, but without the androgen receptor on the maternal-inherited X-chromosome, masculinisation fails and so does male reproduction. Thus, the degenerate Y-chromosome gene content has become specialised to maintain the ancestral dosage of the few remaining homologous XY gene pairs. Moreover, the need for this X-chromosome gene pairing has been critical for the survival of these remaining Y-chromosome genes, and is essential for the segregation of X- and Y-chromosomes during male spermatogenesis. It is also the case that a higher proportion of those genes on the female X-chromosome which do have a Y-linked homologue escape X-inactivation, thereby favouring expression of the maternal copy (Sin & Namekawa, 2013). Thus, males benefit hugely and, indeed, they only maintain their masculinity through inheritance of the mother's X-chromosome. The X-chromosome is also characterised as having a disproportionately high number of genes, which are expressed in the placenta and brain (Graves, 2010). The placenta and brain are two very important organs that have been integral to the success of human evolution and will be dealt with in subsequent chapters. However, it is important to note that the matriline has played a leading role for the co-adapted evolutionary development of these structures (see Chapter 6).

We may, to some extent, gain an overall picture of the functional role played by the sex chromosomes in humans from clinical syndromes. Such experiments of nature include Turner's syndrome in females who are lacking one X-chromosome (XO). Understanding this requires a little more genetic information (Knickmeyer & Davenport, 2011). Females have two X-chromosomes of which one is suppressed by the long non-coding RNA (*Xist*) gene, resulting in the process called X-inactivation (Lee & Barolomei, 2013). Silencing the genes of one X-chromosome in females ensures that the balance for gene dosage is secured across both sexes, as males have only one X-chromosome. The loss of one X-chromosome in females (as in Turner's syndrome) produces a phenotype that differs according to which parent, mother or father, provided the inheritance for their only remaining X-chromosome. Only 3 per cent of pregnancies are viable with XO embryos. Most of these pregnancies terminate very early in the first trimester (Urbach & Benvenisty, 2009) due to failures in placental development. This is due to loss of the maternal X-chromosome. Generally speaking, Turner's syndrome (XO) survivors are usually missing the paternal X-chromosome, which impairs brain functioning, especially in the visual–spatial domain, and in mathematics. Those exceptionally few XO females that do survive placental dysfunctioning, due to the loss of the maternal-inherited X, have lower scores on most social and cognitive measures, and magnetic resonance imaging (MRI) scans show their neocortex to be thicker in the temporal region. Female patients missing the paternal X have enlargement of grey matter in the frontal cortex, but show little evidence of brain dysfunction. Again, we may conclude that it is primarily the contribution of the female X-chromosome which provides the expression of genes to ensure adequate placental development and advanced brain functioning.

Males with an additional X-chromosome (XXY), named Klinefelter's syndrome after its discoverer, possess testes but no sperm are produced. Their brains are also changed, and they develop autistic symptoms with decreased brain activation in areas of the frontal cortex (Viana et al., 2014). These frontal regions of the brain are important to social cognition, and this is revealed in XXY males who are unable to make correct recognition of affect (happiness, sadness, anger) from the examination of facial expressions. It is, therefore, not simply those genes functionally engaged in female and male reproduction that have accumulated on the female X-chromosome, but also those genes in the brain that are especially involved with social and intellectual abilities.

New gene-sequencing technologies have recently identified certain gene mutations associated with X-linked intellectual disability. These pathological variant genes have been identified in those females with skewed X-inactivation, resulting in their X-chromosome genes being expressed only when they are inherited from the father's X-chromosome. There is also a male over female predominance for Autism Spectrum disorders (ASDs) according to which of the X-chromosome's genes are expressed (Hoffbuhr et al., 2002). ASDs comprise a complex group of behaviourally related disorders that are primarily found in males and are genetic in origin. One of the genes that has been identified on the X-chromosome that is important in ASD is the gene responsible for DNA methylation (*MeCp2*). This gene is required for silencing other genes during brain development, thereby regulating the spatial and temporal expression of developmentally important genes. Specific mutations to the *MeCP2* gene results in reduced brain size (primarily cortex and cerebellum) producing impaired social interactions (Hoffbuhr et al., 2002). These so-called Rett syndrome patients share many of the neurological symptoms of autistics, while specific *MeCp2* mutations have been further identified in association with neuropsychiatric disorders, such as schizophrenia. Skewing of X-inactivation in favour of the healthy X-chromosome avoids the *MeCp2* mutation and hence expression of the syndrome in females, but this is not possible for males, who have only the one X-chromosome.

Returning to the sex steroid hormones and their influence on behaviour, pioneering work on small-brained rodents (rats and mice) in the 1960s identified a critical period for sexual differentiation of the male brain around the time of birth (reviewed in de Vries & Sodersten, 2009). A single injection of testosterone to female rat pups at this critical time is sufficient to depress their feminine sexual behaviour later in life. Moreover, if at this later stage their ovaries are removed, and this neonatal androgenised female is now given the male hormone testosterone, then there is an adult enhancement of male patterns of behaviour. These experiments led to the conclusion that late in the embryonic life of males, the central nervous system is sexually undifferentiated, and future masculine behaviour is organised by hormones secreted by the immature testes (Herbert, 2015). These studies provided a biological basis for explaining sexual dimorphic behaviour, and it was tempting to relate such findings to humans. However, the rat brain has little resemblance to the human or any primate brain, and when similar experimental studies were undertaken with monkeys that possess a much larger executive brain than rodents, the outcome was very different. When female monkeys were prenatally treated with the hormone testosterone, they continued to show female patterns of behaviour when they reached maturity and were paired with males, albeit at considerably reduced levels (Herbert 2015). Importantly, these females did not exhibit masculinised behaviour, and they were delayed in starting their ovulatory menstrual cycles. In this context, the monkey's brain is very different when compared to the rodent, and a significant effect of social rearing is found which appears to be more consequential. Early social experiences can themselves have a major influence on the development of sexual behaviour in monkeys. This ranges from complete sexual inadequacy resulting from early separation of the male infant from the mother, and differences in assertive 'rough and tumble' play behaviour. The same applies for male monkeys that, after weaning, are housed and raised in isolation from females (Kraemer, 1997). Such isolation also results in a decline of their adult male mating behaviour with females. Thus, the large brain found in monkeys of both sexes has provided a degree of emancipation from hormonal determinants of sexual behaviour. This emphasises the importance for consideration of social learning in the context for which their long-developing and complex executive brain matures.

Evidence for early influences of hormones on sexual differentiation of the human brain has been revealed from clinical findings, notably from the medically named 'adrenogenital syndrome' (AGS) (Bancroft, 2002). In AGS females, the syndrome results in them being exposed to testosterone as a result of pathological over-activity of their foetal adrenal glands. The foetal adrenal glands normally produce low levels of testosterone in both males and females. Overactivity of these adrenal glands in human females results in these females being born with masculinised external genitalia and, until relatively recent years, they were mistakenly raised as boys. Today, appropriate therapy has prevented any postnatal over-production of testosterone and this early recognition of the syndrome no longer results in mistakes of gender assignment. When given appropriate treatment at birth, such AGS females clinically resemble the experimental group of monkeys that were masculinised by testosterone, which was administered when they were developing *in utero*. Behaviourally, these testosterone-exposed girls show more energetic play activity (tomboys), but when they reach maturity, there is little evidence to suggest that their sexual proclivities are notably different from most other females. In the main, their sexual orientation is that of a female; they prefer men, get married, and have children. However, there have been recent reports of gender dysphoria and bisexuality in conjunction with the 'tomboy' status. Clearly, then, early exposure to testosterone *in utero* does not have the same irreversible consequences for the sexual development of the human female brain as that which is found with the experimental testosterone treatment of rodents.

The same is not, however, true for girls with the AGS who are not treated at birth, but are reared as boys and continue to be exposed to very high levels of testosterone into puberty (Leroy, 2007). These girls have a masculine appearance, a sexual preference for other girls, and usually wish to abandon their genetic sex and be formally reassigned as males. Might this finding point to an important influence of gender-specific rearing of humans for their future sexual orientation? Categorically, the answer is no, or yes, depending on how the evidence is interpreted. Somewhere between these two groups of children, (those females with the syndrome treated or untreated) is a condition that was first reported for an isolated population of families in the Dominican Republic. A deficiency in a recessive gene carried by their mother produced a condition in male offspring whereby insufficient amounts of another hormone, dihydrotestosterone, are metabolised from this male's testosterone. As a consequence of these low levels of dihydrotestosterone, the external genitalia are not fully masculinised at birth, and boys with this inherited condition (named locally as *guevodoces*) are initially reared as girls (Pardridge et al., 1982). At puberty, their increased production of testosterone is sufficient to overcome the enzyme deficiency and masculinisation proceeds, together with a reassignment of psychosexual identity to that of being male. Their clitoris enlarges to form a penis, their testes descend into the labia, and their body hair assumes the appearance of males. At face value, this finding would suggest that early exposure of the brain to testosterone appears to overcome the female sexual identity to which these *guevodoce* children were assigned and reared as infants. Such a view is not, however, without its critics, as the gender identity of these boys is not entirely unambiguous. Their collective name *guevodoces* (literally translated as 'eggs at 12'; the eggs being a colloquial term for testes) implies a special place for these children in the eyes of their community, and the children themselves are already aware of their special status by the age of 6–7. Normal levels of testosterone at puberty, and a social status that recognises their masculine predisposition, is sufficient to ensure their future role as males.

The clinical equivalent of early castration in human males can be found in Tfm, which occurs in genetic males with physiologically normal levels of circulatory androgens but with target tissues that are insensitive and do not respond to these hormones. As mentioned earlier, these subjects have genitalia that are unambiguously female at birth, and they are subsequently reared as girls. At puberty, masculinisation fails to occur, and these genetic males develop breasts and have all the appearances of normal women except that they have scant pubic hair. Psychosexually, their orientation is directed towards men, they often marry and, although infertile, show normal maternal behaviour towards adopted children.

Thus, although testosterone plays a primary and significant role in mammalian masculinity, there is no single gene which directly codes for this outcome. Moreover, testosterone's effects on the development of brain and behaviour are primarily dependent on the nuclear androgen receptor (see Chapter 3 for a detailed account of epigenetics). These nuclear effects of testosterone are only made possible and brought about via the androgen receptor, for which there most certainly is a gene, but this gene is provided by the female X-chromosome which males inherit from the mother. Many human societies have placed a premium on having male offspring, and we already have an example of this with the *guevodoce* children of the Dominican Republic (Pardridge *et al.*, 1982). This special *guevodoce* status implies a male achieving masculinity at 12 years is better than no male at all. Why so many societies embrace males with a leadership role will be dealt with later, but it is clearly female genetics that are integral to the male's reproductive status. Moreover, mammalian reproduction is only successful through implantation and viviparity, with the female sustaining a viable placenta, a warm secluded postnatal environment, lactation and devoted maternal care. The Y-chromosome, which is integral to the determination of the male sex, plays no part in these vital viviparous reproductive events. It might therefore be argued that it is primarily by the grace of the female's genome that males are created (see Chapter 3 for the origins of the male *SRY* sex-determining gene).

To briefly summarise, mammalian evolution has seen a progressive attrition of genes on the male sex determining Y-chromosome, leaving available only those genes that are specialised for male reproduction. The Y-chromosome has undergone considerable degeneration and does not have a matching Y-chromosome partner. The solitary Y-chromosome has consequently shown reduced recombination with the X-chromosome partner over most of its length. This has resulted in the reduced ability for natural selection to act on the Y-chromosome. Moreover, gene decay on the non-recombining Y-chromosome is a consequence of the accumulation of deleterious mutations and the absence of mismatch repair enzymes. From an evolutionary viewpoint, selection pressures for change acting on the few remaining duplicate X and Y genes can no longer be advantageous just to males, while those genes which are unique to the Y-chromosome have been neutralised for everything except the male sex-determining genes. Even the male unique *SRY* sex-determining gene is a hybrid of the autosomal *Dgcr8* and *Sox3* genes which are expressed in both parents. It is, of course, of interest to both sexes that males are fertile and sexually active, so it is highly unlikely that the Y-chromosome will ever completely disappear from humans as indeed has happened in certain rodent species (Kuroiwa et al., 2011).

Most of the advantageous innovations that have accrued during mammalian evolution have come about primarily as a consequence of in-utero development and viviparity. Pregnancy, placentation and maternal care have each played an integral role in the success of mammalian evolution, and these innovations, although maternally led, are as much in the father's interest as that of the mother. Offspring of both sexes benefit from viviparity in mammals. However, the theory that female sex determination represents a 'default' state is far removed from what we now understand about the developmental differentiation of the female and male reproductive systems. Indeed, developmental differentiation of the female reproductive system is an active process involving genes across many chromosomes, some of which promote ovarian development, uterine development and development of female genitalia. Mammalian sexual development at the genetic level can be considered 'as a story of male and female crucial alliances and opposing forces' (Eid *et al.*, 2015), but, nevertheless, a process in which the female undeniably plays a leading role. The timing for gene expression is of crucial importance and even a minor delay in male *SRY* expression can result in a failure of masculinity produced by XY sex reversal. Although the male Y-chromosome is specialised for male reproduction, other genes of bi-parental origin are also integral to this process.

Until recent years, factors involved in male sex determination have received considerably more attention than was given to the sex determination of females, possibly because being female was thought to be the male default state. It is only in recent years that the genes which play a distinct role in female ovarian development have been identified (Biason-Lauber & Chabolissier, 2015). Prior to the developmental onset of sex determination, those precursor somatic cells which become committed to producing the ovary already express more genes than those early somatic cells which are the precursor for male testes development. This would suggest that very early somatic gene lineages are already primed with a bias toward female fate. Ovarian development is very much an active process and certain genes involved with this process (*WNT4*) have been identified which are also known to suppress sexual differentiation in the male. Moreover, XX individuals have been identified who develop testes in the absence of the *SRY* male sex-determining gene. This leads to a conclusion that the developing XX female germline expresses some factors which have both anti-testes and pro-ovarian developmental capabilities.

Clearly, mammalian sex differences are not as simple and straightforward as we used to think. Gone is the time when we might consider the origin of sex differences as simple categorical processes.

We now know that, at the genetic level, sex differences are not simply a case of 'presence' or 'absence' for expression of the male *SRY* gene. A more realistic viewpoint would support the conclusion that developmental sex difference are represented by 'crucial alliances and opposing forces' across both the male and female genomes (Biason-Lauber & Chabolissier, 2015). However, when considered from an evolutionary perspective, it can be concluded that it was selection pressures operating primarily via the matriline and the experience of in-utero viviparity which assumed the leading role in shaping mammalian sex differences. From a biological and mechanistic perspective, female sexual development is a very active process, far removed from being a biologically defined 'default state'.

REFERENCES

Bachtrog, D. (2013). Y chromosome evolution: merging insights into processes of Y chromosome degeneration. *Nat. Rev. Genet.* 14: 113–24.

Bancroft, J. (2002). Biological factors in human sexuality. *J. Sex. Res.* 39: 15–21.

Biason-Lauber, A. & Chabolissier, M. C. (2015). Ovarian development and disease: the known and the unexpected. *Semin. Cell Dev. Biol.* 45: 59–67.

Bion, S. & Toniolo, D. (2000). X chromosome genes and premature ovarian failure. *Semin. Reprod. Med.* 18: 51–57.

Cortez, D., Marin, R., Toledo-Flores, D., *et al.* (2014). Origins and functional evolution of Y chromosomes across mammals. *Nature* 508: 488–93.

de Vries, J. & Sodersten, P. (2009). Sex differences in the brain: the relation between structure and function. *Horm. Behav.* 55: 589–96.

Eid, W., Potz, L. & Biason-Lauber, A. (2015). Genome-wide identification of *CBX2* targets: insights in the human sexual development network. *Mol. Endocrinol.* 29: 247–57.

Ferguson-Smith, M. A. & Rens, W. (2010). The unique sex chromosome system in platypus and echidna. *Genetika* 46: 1314–19.

Graves, J. A. (2010). Review: Sex chromosome evolution and the expression of sex-specific genes in the placenta. *Placenta* 31: S27–32.

Herbert, J. (2015). *Testosterone: Sex, Power and the Will to Win*. New York, NY: Oxford University Press.

Hoffbuhr, K. C., Moses, L. M., Jerdonek, M. A. *et al.* (2002). Associations between *MeCP2* mutations, X-chromosome inactivation, and phenotype. *Ment. Retard. Dev. Disavil. Res. Rev.* 8: 99–105.

Jiminez, R., Barrioneuvo, F. J. & Burgoa, M. (2013). Natural exceptions to normal gonad development in mammals. *Sex Dev.* 7: 147–62.

Katoh-Fukui, Y., Miyabayashi, K., Komatsu, T., et al. (2012). Cbx2, a polycomb group gene, is required for *Sry* gene expression in mice. *Endocrinology* 153: 913–24.

Knickmeyer, R. C. & Davenport, M. (2011). Turner syndrome and sexual differentiation of the brain: implications for understanding male-biased neurodevelopmental disorders. *J. Nueurodev. Disord.* 2011: 293–306.

Kraemer, G. W. (1997). Psychbiology of early social attachment in rhesus monkeys. *Ann. N.Y. Acad. Sci.* 807: 401–18.

Kuroiwa, A., Handa, S., Nishiyama, C., et al. (2011). Additional copies of *CBX2* in the genomes of males and mammals lacking *SRY*, the Amami spiny rat (*Tokudaia osemensis*) and the Tokunoshima spiny rat (*Tokudaia tokunoshimensis*). *Chromosome Res.* 19: 635–44.

Lee, J. T. & Barolomei, M. S. (2013). X-inactivation, imprinting, and long noncoding RNAs in helath and disease. *Cell* 152: 1308–23.

Leroy, F. (2007). [Hermaphrodites and intersexed individuals from myth to reality. Second part]. *Vesalius* 13: 77–81.

Matsumoto, Y., Buemio, A., Vafaee, M., et al. (2013). Epigenetic control of gonadal aromatase (cyp19a1) in temperature-dependent sex determination of red-eared slider turtles. *PLoS ONE* 8: e63599.

Migeon, B. R., Brown, T. R., Axelman, J., et al. (1981). Studies of the locus for androgen receptor: localization on the human X chromosome and evidence for homology with the Tfm locus in the mouse. *Proc. Natl Acad. Sci. USA* 78: 6399–43.

Pardridge, W. M., Gorski, R. A., Lippe, B. M., et al. (1982). Androgens and sexual behavior. *Ann. Intern Med.* 96: 488–501.

Sin, H. S. & Namekawa, S. H. (2013). The great escape: active genes on inactive sex chromosomes and their evolutionary implications. *Epigenetics* 8: 887–92.

Sun, S. & Heitman, J. (2012). Should Y stay or should Y go: the evolution of non-recombining sex chromosomes. *BioEssays* 34: 938–42.

Tardat, M., Albert, M., Kunzmann, R., et al. (2015). *Cbx3* targets *PCR1* to constitutive heterochromatin in mouse zygotes in a parent-of-origin-dependent manner. *Mol. Cell* 58: 157–71.

Urbach, A. & Benvenisty, N. (2009). Studying early lethality of 45,XO (Turner's Syndrome) embryos using human embryonic stem cells. *PLoS ONE* 4: e4175.

Viana, J., Pidsley, R., Troakes, C., et al. (2014). Epigenomic and transcriptomic signatures of a Klinefelter syndrome (47,XXY) karyotype in the brain. *Epigenetics* 9: 587–99.

Wang, Z., Say, Y. L., Zhang, J., *et al.* (2014). Complete androgen insensitivity syndrome in juveniles and adults with female phenotypes. *J. Obset. Gynaecol. Res.* 40: 2044–55.

Yamauchi, Y., Riel, J. M., Stoytcheva, Z., *et al.* (2014). Two genes can replace the entire Y chromosome for assisted reproduction in the mouse. *Science* 343: 69–72.

2 Epigenetics: The Gene Environment Interface

The discovery of DNA's double-helix structure from which genes are constructed and the heritability this endows on organisms won Francis Crick, James Watson and Maurice Wilkins the Nobel Prize for Physiology or Medicine awarded in 1962. It is generally acknowledged that the crystallographic work of Rosalind Franklin on DNA structure was integral to this discovery, but she unfortunately died of cancer in 1958, and Nobel Prizes are never awarded posthumously. These findings were followed in 1965 by a further Nobel Prize to Francois Jacob, Jacques Monod and Andre Lwoff for the discovery of how the structural information contained in the gene's DNA is translated for production of a messenger RNA. RNA represents one strand of the gene's DNA coding sequence and carries these instructions to the ribosomes, the cell's construction factory for the building of proteins. It is the nucleotide sequences of DNA which contain information on the type of amino acid building blocks that go to make a protein, but it is the messenger RNA that aligns the code for these building blocks to construct that protein. This advance from DNA structure to function, underpinning the nature and nurture of genetics, laid the foundations for molecular biology as we know it today (Lewin, 2006). The Watson and Crick model of gene structure represented a coded message containing two types of information (gene structure and heritability) such that when a cell divides, each daughter cell receives the exact same copy of the parent gene. Thus the nature of humankind is embedded in a genetic code for each of us; a code that is transmitted to future generations. Nurture provides the sum total of how our individual physical and social environments shape the expression of this code throughout our lifetime development.

Some 15 years ago the human genome was sequenced and approximately 20,000 genes were identified that encoded for the protein building blocks in each and every cell. Considering it was originally believed that the human genome would contain millions of genes, this significant reduction from what was expected came as something of a surprise. Comparative studies across the genomes of widely differing species found very little correlation in the size of these genomes with organism size. As an example, the relatively primitive salamander has a genome some 15-times larger than that of humans (Gall, 1981). This paradox was thought to be resolved when it was found that most of the genome does not code for genes. Indeed, the 20,000 identified genes represented a mere 1.5 per cent of total DNA and the rest was labelled as 'junk DNA' or 'non-coding DNA' (Ohno, 1972). Junk DNA was thought to be an integral part of those DNA sequences that are remnants of our past evolutionary history. Sydney Brenner, the molecular biologist and Nobel prize winner, drew attention to the usefulness of the term 'junk', which you store in the attic in case it may serve a useful purpose one day in the future. Naming the non-coding DNA as junk does not mean this DNA is entirely useless, and it has been hypothesised to serve as a reservoir for future evolutionary innovations (Comings, 1972). One such innovation which is now receiving considerable attention is the regulation of gene expression by non-coding RNA. Perhaps the best known of these long non-coding RNAs is Xist RNA, which coats the X-chromosome and acts as a scaffold to recruit DNA-silencing factors such as the 'polycomb' repressive complex (Zhao et al., 2008).

The extraordinary amount of non-coding DNA is made up of simple DNA repeat sequences, transposable elements and pseudogenes (de Koning et al., 2011). It is now known that embedded in these non-coding sequences is the DNA which determines where, when and how the expression of the coding DNA sequence is itself regulated. Such gene regulation is determined by the spatio-temporal, highly specific transcriptome, which codes for the transcribing of

DNA. Gene transcription factors bind close to the protein coding regions of DNA in a combinatorial fashion to specify the on-and-off state of specific genes. These transcription factors bind in distinct combinations and may even bind differentially to mother's or father's parental allelic copy of the gene.

Not all genes are active at any one moment in time, and although transcription factors are necessary for bringing about gene expression, genes may also be actively silenced (chemically DNA-methylated) and thereby not available for transcription. Those regions of DNA which serve as promoter activator switches for DNA gene expression are rich in nucleotide sequences (so-called DNA CpGs, of which there are some 29 million in the human genome). When these nucleotide CpG sequences are methylated, they stabilise the 'inactive' state of a gene's DNA coding sequence (Bird, 2007). Thus, gene inactivation is a positive state of silencing, just as gene activation represents a positive state for promoting gene expression. Both activation and silencing of genes are integral to formation of the phenotype.

Although it was Waddington who first introduced the term 'epigenetics' as a descriptor linking the phenotype with genotype during development (Waddington, 1946), it was some decades before this feature of heritability became linked to DNA 'memory' marks (DNA methylation and demethylation). Other modifications to the histone chromatin exoskeleton that supports DNA also register, signal or perpetuate the activity state of genes via structural changes of the chromatin protein framework. It is principally these two types of epigenetic mark on DNA, methylation and demethylation, together with changes to the histone protein framework that regulate DNA's gene expression capabilities. Methylation marks applied directly to DNA primarily result in long-term gene silencing. Those methylation marks which are applied to the histone protein exoskeleton of DNA regulate access of transcription factors required for generating the gene's expression programmes. DNA is wound around a complex of four different histones, each of which

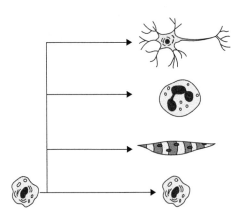

FIGURE 2.1 Cell differentiation: an epigenetic process. Germ cells have the capacity to develop into many cell types (200 in all), but all possessing the same genome.

can be modified (methylated, acetylated) thereby allowing access of gene promoters to DNA (switch-on) or restricting such access (gene silencing) (Bird, 2007). The patterns of histone modifications, sometimes referred to as the histone code, are sufficiently diverse as to provide for the huge complexity and specificity of gene expression. There are many more chemical marks that can be applied to histones, some of which change the three-dimensional structure of DNA, thereby allowing for DNA interactions both within the same chromosome and across different chromosomes. One distinct advantage of this kind of epigenetic regulation for gene transcription is that different cell types, containing exactly the same genomic DNA, can be developmentally regulated in different ways (Figure 2.1). Indeed, histone modifications underpin a code of their own, the histone code of gene transcription. However, transcription is not even this simple, and during development some gene promoters may have both activating and repressive histone marks available simultaneously, thereby providing a potential time frame for a given gene's expression. The most recent versions of the human genome have revealed multiple tiers of gene regulatory networks which, if they could be printed out, would leave a paper trail stretching some 30 km or a stack of A4 sheets of paper some 16 m high. Among DNA's three billion nucleotide letters, the regions that code for proteins only represent 1 per cent of the genome (20,000 genes). The rest of these nucleotides make up an unprecedented number of functional regulatory elements including enhancers, promoters, silencers, insulators and locus control regions, with 2.9 million different regions of chromatin that are available for accessing DNA complexes, such as PRC1 and PRC2, which interact with a large number of Lnc-RNAs.

The structure of RNA is chemically similar to that of the DNA which provides the template for its synthesis. However, RNA is single-stranded and exists in three different forms: ribosomal RNA, messenger RNA and transfer RNA. Recently there has been considerable interest in a form of non-coding RNA that has been found to play a role in nullifying the action of messenger RNA. These long non-coding RNAs (Lnc-RNA), of which some 9000 have been catalogued, also function to regulate gene expression by modifying the structure of chromatin, while other short interfering RNAs mediate post-transcriptional gene silencing by causing RNA degradation. Still other, so-called micro-RNAs, fine-tune gene expression by targeting messenger RNA (Heard & Martienssen, 2014). Lnc-RNAs also play an important role in epigenetics, acting as a scaffold and tether for the recruitment of polycomb proteins that regulate chromatin state (Kung *et al.*, 2013). Chromatin polycomb complexes such as the gene *Ezh2* in PRC2 is a key to activating histone H3K27 methyltransferase. The gene *Cbx2* in PRC1 is important for stimulating the genetic pathway that determines male testes development, and inhibits sexual differentiation of the ovaries (Biason-Lauber & Chabolissier, 2015). The Lnc-RNA named HOTAIR guides chromatin-modifying complexes (PRC2) to target gene promoters, leading to gene silencing. The steroid hormone oestrogen is engaged with transcriptional regulation of HOTAIR and is associated with breast cancer (Bhan &

Mandal, 2016). Because Lnc-RNAs are attached to chromatin during transcription and are transcribed from a single locus in the genome, they have the capacity to direct locus-specific silencing for a single allele. In this way they provide a further dimension to locus-specific control, a dimension which cannot be undertaken by protein transcription factors alone. Transcription factors do not recognise DNA's positional memory once they are translated. However, when attached to chromatin, the length of Lnc-RNAs permits them to 'reach out' and recruit additional epigenetic modifying complexes.

In addition to regulating gene expression by recruiting epigenetic complexes, Lnc-RNAs can also directly influence transcription by competing for transcription factor binding. Lnc-RNAs are further involved in processing and providing stability control of mRNA, and can serve as host genes for small s-RNAs. Many Lnc-RNAs are translated into small peptides, but whether these peptides are functional or just translational noise still remains to be determined. Should these small peptides have no meaningful function, then they may either be eliminated by natural selection, or remain and serve as a reservoir for the future evolution of adaptive functions (Kung *et al.*, 2013).

Non-coding RNAs may themselves be subjected to epigenetic regulatory control by the M6A demethylase enzyme. One of the important genes that activate mRNA demethylation is *FTO* (Fat Mass Obesity Gene), which is expressed in fatty adipose tissue, but also in the brain. Insulin has been identified as a regulator for *FTO* expression. Genetic variants of this gene have been identified in Alzheimer's patients, and the reduced executive brain size of Alzheimer's patients during ageing is thought to be due to variant *FTO* risk alleles (Reitz *et al.*, 2012). Carriers of the *FTO* risk alleles also have a wide variety of differentially methylated (silenced) sites in the genome that regulate fat cell development, as well as regulating the feeding and appetite gene (*NPY*). The RNA (M6A) demethylase enzyme is also known to be influenced by insulin, which in turn is itself affected by stress, providing a clear pathway for linking the environment to certain pathological phenotypes. The epigenome is thus represented by chemical marks that regulate DNA and its scaffold, informing the cell on which gene to express and which to keep silent. In this way, cells with identical DNA can develop into a multitude of tissue types. On RNA alone, more than 100 different types of mark are found that control gene expression as part of the 'epi-transcriptome'. Similar to the epigenetic marks on histones and DNA, RNA methylation is reversible. It is known that more than 12,000 methylated sites on RNA originate from 7000 genes. The organisation of these RNA methylated marks allows cells to produce multiple versions of a protein from a single cell, thereby providing a fundamental role in the differentiation of various cell types. Just like non-coding RNAs can synchronise the activity of multiple transcripts, it is thought that RNA methylation marks may play a similar role.

The understanding of molecular genetics is gaining in complexity by the month, and the popular view of 'a gene for this function, or a gene for that function' is very far from the reality of the complex transcriptional network programmes that are available for regulation of gene expression. Moreover, there are no genes that function in isolation from other genes. It is thus more appropriate to think of a gene's action as part of a functional network, with some genes being common to multiple networks and others (the genetic hub) which play a leading role in a given network's regulation. How these different tiers of gene regulation (activation or repression) are deployed represents this field of 'epigenetics', a field which is fundamental to understanding development and, importantly, how development itself may be influenced by the environment. These multiple levels for control of gene expression also point to far-reaching influences on developmental disorders. Such detrimental influences extend far beyond the established mutations and polymorphisms that are currently well recognised as an integral part of heritable dysfunctions.

The magnitude of epigenetic regulatory modifications to the genome which occur in the lifetime of individuals are mainly advantageous to that individual, enabling adaptation to their current environment. However, there are some environments, particularly those for which mankind has been instrumental in creating, that are not advantageous (e.g. warfare, famine, child abuse, social neglect). Such aversive stressful environments can create relatively long-term pathological changes to DNA function. Fortunately, most of these maladaptive, epigenetic-induced changes to gene expression are not transgenerationally heritable due to the germline (egg and sperm) reprogramming of DNA's epigenetic marks. However, if these maladaptive changes are sufficiently devastating for their present generation, then their impact on the brain's development of the present generation may in turn produce adverse social interactions that are maladaptive towards the next generation.

Germline reprogramming enables genes to achieve totipotency, a process that is required for normal development in each successive generation (Surani et al., 2007). As mentioned earlier, most of the key regulatory genes that are engaged specifically in development become permanently methylated and silenced once they have performed their developmental tasks. It is these methylation marks on DNA, together with the histone protein modifications, that can bring about selective gene silencing. Hence, it is this lifetime of gene silencing which needs to be reversed in order for the development of the next generation to proceed. Erasure of past epigenetic silencing marks from the genome is an essential part of resetting the germline's developmental programme to its original totipotent state. In this way, pre-existing gene expression patterns are erased from the germline, and are subsequently re-established according to their male or female germ cell origins, thereby ensuring successful development of the next generation (Surani et al., 2007). Some regulatory elements, such as retro-transposed viral elements that are potentially hazardous, are able to evade this demethylation silencing. However, many retro-transposed elements have actually become safely incorporated into the genome, and are adaptively engaged in gene regulation (e.g. a number of the imprinted genes which have imprint control regions of retroviral origin). Indeed, it is the repeat sequences located in retroviral DNA that are thought to have driven cell fate determination in the evolution of the mammalian placenta (Wu *et al.*, 2016).

Thus, at the very earliest phase for development of the fertilised egg, DNA methylation profiles can be found that are specific to each of the distinct future cell lineages, namely those going to form the placenta and those going to form the embryo itself. These tissue-dependent regions of the early blastocyst are initially determined by histone methylation, which functions as a 'landmark', responsible for setting up the future cell lineage-specific DNA methylation profiles. Because such early landmarks are themselves derived from the repeat sequences located in retroviruses, this would suggest that 'foreign' DNA sequences may have helped to drive cell fate determination in the evolution of placental mammals.

RETROTRANSPOSONS

Retrotransposons are selfish genetic elements that have been used to drive evolution but which, if not silenced, may also create serious errors in neural development (Rett syndrome) and somatic development (cancer). The evolutionary impact of retrotransposons has depended on their ability to regulate gene expression through their predisposition for DNA methylation silencing. This has further required engagement of other germline genes (*Trim28* and *Zfp57*) for the specific silencing of retrotransposons, and safeguarding transcriptional dynamics by controlling DNA stability (Turelli, 2014).

A number of studies in mammals have found an association between phenotypes and the silencing of transposable elements, an example of which is the Agouti mouse (Daxinger & Whitelaw, 2012). A retrotransposon insertion upstream of the Agouti gene leads to yellow fur, obesity and diabetes. Incomplete erasure of the phenotype occurs on passage through the matriline, but not the patriline, and is due to incomplete erasure of methylation in the female germline. The oocytes of the female germline are also able to develop aneuploidy defects, and embryonic lethality due to the elevation of L1 transposons. Such oocyte attrition has been suggested to select for oocyte stability and for those oocytes that are best suited for normal development of the next generation (Malki et al., 2014).

The brain is an important target for retrotransposon mobilization, and L1 sequences are usually transcriptionally repressed by germline methylation. However, L1 transposons are initially hypomethylated in the developing brain, a condition which is subsequently repressed by Sox2, and transcriptionally silenced by DNA methylation. L1 retrotransposition tends to occur in the later phases of neurogenesis, thereby inducing neural cell mosaicisms which underpin genetic diversity across neural sub-populations. Transposons may also result in the neural disorders of schizophrenia and the neurodevelopmental Retts syndrome, due to the misregulation of these mobile elements (Coufal et al., 2011). Thankfully, the germline insertions of L1 retrotransposons are rarely found in those regions of the genome where they could produce a deleterious phenotype. This is thought to have been avoided as a consequence of strong selection pressures against the occurrence of such mutations during evolution (Erwin et al., 2014). However, neural progenitor cells that are derived from patients with Retts syndrome and are carrying mutations to MeCp2, the DNA methylating protein, have been shown to exhibit increased susceptibility to L1 retrotransposition. Epigenetic methylation silencing of retrotransposed DNA is therefore as integral to ensuring normal brain development as is the reversal of DNA methylation for the developmental gene transcription.

REPROGRAMMING THE EPIGENOME

The developmental origins of males and females commence through their germline. Development is initiated via the production of primordial germ cells, the precursors for sperm and for oocytes. In order for these male and female germ cells to produce the next multicellular generation, they first require their own DNA to be reprogrammed by the removal of the silencing methylation marks. All of the various epigenetic DNA marks which are specific to cell and tissue types, to genomic imprints and to X-inactivation need first to be removed.

These marks are then subsequently reset at various stages throughout the next generation's early development. Indeed, there is a genetic programme responsible for reprogramming these DNA epigenetic marks, a programme which is fundamental to the recreation of all mammalian life forms (Nakagawa & Yamanaka, 2010). Interestingly, the knowledge and understanding of this process has been applied to the reprogramming of 'stem cells', an important growth area in biological and medical sciences, which may one day be routinely used for replacement or repair of damaged tissues. Remember 'Dolly' the sheep? She was created from the reprogramming of a mammary gland cell and so named after a famous singer who is well-endowed in that region of her anatomy. Male chauvinism raising its head in science once again! More seriously, the Nobel Prize was awarded to John Gurdon and Shinya Yamanaka in 2012 for their pioneering studies in this area of genetic reprogramming and its importance for stem cell research (Gurdon, 2013).

Epigenetic differences already exist between male and female genomes at the very earliest stage of germline development, especially with respect to the timing of reprogramming. Reprogramming of the female germline occurs in the oocytes of the developing foetal ovary, thereby providing totipotency for the next generation. In effect, this means that the origins of my mother's genome for my foetal development underwent reprogramming within my grandmother's womb, two generations ahead of its deployment in myself. Such matrilineal forward planning may account for the puzzling findings of low birthweight babies seen in the second generation of children born many years after the Dutch famine of 1944 (El Najj et al., 2014). It is notable that there are relatively few maternal oocytes (eggs) produced in a human female's reproductive lifetime, especially in comparison with the millions of sperm that males produce. Moreover, the mother's oocytes need to sustain their genetic stability for many years before fertilisation occurs, whereas the lifetime of an active motile sperm lasts but a few hours. Unlike the maternal oocytes, the male sperm has to undergo multiple cell divisions following their initial round of reprogramming in the sperm's paternal germline. This introduces instability into the genetic life history of males, as multiple replications of DNA during sperm production generate many more genetic errors than occurs in female oocyte production (Marchetti et al., 2007). Hence, a second phase of reprogramming for the male genome is required post-fertilisation, and this takes place at the very earliest stage of the next generation's development, during the period when the fertilised egg develops to form the zygote (Jenkins & Carrell, 2012). This is also a phase, indeed the earliest stage possible, when the strands of male and female DNA become aligned to permit mismatch repair of paternal DNA, using the stable, reprogrammed copy of maternal DNA as the template (Wossido et al., 2010).

Why is reprogramming of the germline so important and are there any differences, other than timing, between reprogramming of the mother's germline compared to that of the father? At fertilisation, the male genome is tightly packaged in proteins (protamines) in order to maximise miniaturisation of sperm for their incredibly long swim to the egg. However, these protamines need to be removed at fertilisation in order to permit transcriptional access to the paternal genome, thereby enabling male genes to be expressed after fertilisation. Thus, following fertilisation, the female genome is responsible for releasing the male genome from its protamine cage and ensuring its replacement with the transcriptionally flexible histone proteins. Without this new histone exoskeleton for male DNA, there would be no gene expression, and no access available for the enzymes required to ensure the second (post-fertilisation) round of epigenetic reprogramming undertaken to enable repair of the male genome. More importantly, it is the maternal genome that is responsible for producing the proteins (demethylation enzymes) that are required for this second phase of paternal genome reprogramming, and also ensuring that the paternal DNA matches to that of the stable maternal genome (Hatanake et al., 2013). Several reports have shown that the paternal genome undertakes post-fertilisation demethylation reprogramming before replicating its DNA, whereas the maternal genome, over this same time period, retains DNA silencing (methylation) at a constant level of stability. Thus, maternal proteins in the fertilised egg (zygote) are engaged in the reprogramming of the paternal genome, while at the same time protecting the already reprogrammed maternal imprints from demethylation (Li et al., 2008). This protection is essential for sustaining the genomic imprints (see Chapter 3) and thus maintaining their capacity for expression of the autosomal (non-sex) genes according to their parent of origin in this next generation. In addition to regulating epigenetic stability, the zygotes' maternal proteins also ensure it is their maternal copy of the gene that initiates the first step in development of the placenta (Jedrusik et al., 2010). As if to emphasise this maternal regulatory control over the paternal genome, the sperm's mitochondria, the source of energy production in all cells, are excluded from the fertilised egg, having become energetically 'exhausted' from energising the sperm's marathon swim to reach the egg. Only the maternal mitochondria are inherited by the next generation.

In summary, in order to recreate the next generation, the lifetime's epigenetic clock needs resetting, and the epigenetic marks that have accumulated in the previous generation need to be removed in the developing germline. In this way, the germline achieves a totipotency that is essential for development of the next generation. The process of reprogramming germ cells, although mechanistically similar in males and females, differs in the timing. Importantly, it is the mother's genome which takes the controlling lead in accomplishing this process for the paternal germline, especially during this crucially important post-fertilisation, zygotic phase of development.

A recent finding which may be of importance for intergenerational epigenetic inheritance has been the discovery of certain genes and DNA sequences which escape germline DNA demethylation (Tang et al., 2016). A number of these escapee sequences are retrotransposons, viral DNA which is potentially dangerous and needs to be maintained transcriptionally silent. Retrotransposons are selfish genetic elements that may be used to drive evolution, but which may also create serious errors in brain development (Retts syndrome) and somatic development (cancer). Moreover, there are also a substantial number of these 'escapee' genes which are expressed in the brain, and play a necessary part in the development of the brain for this next generation (Tang et al., 2016). Because most of human brain development takes place postnatally, and because the developing brain has its own genetic reprogramming enzymes, these methylation-silenced 'escapee' genes may represent a means of protecting brain-specific genes from any premature epigenetic modifications. Thus, retention of this early embryonic methylation silencing that is specifically for future brain expression genes protects them from genetic modifications during the earlier stages of embryo development. In this way their functional expression is specifically reserved for the brain, even into the later postnatal stages of its development. Such early conservation silencing of brain genes may also reserve these genes for later selection pressures that are specific to brain function. Hence, some of the 'escapee' genes, notably those which are primarily concerned with developmental growth of the brain, are unique to humans and may escape early demethylation in humans but not in mice. Retention of methylation silencing for such brain-specific genes protects them from epigenetic modifications during the earliest stages of embryo development, thereby conserving their functional expression specifically for the brain. One potential outcome of later genetic reprogramming of brain-expressed genes ensures their reservation for later epigenetic modifications that may be beneficial. Conservation of genes specifically for reprogramming in the brain could thereby permit a more flexible, environmentally responsive, epigenetic brain phenotype than is possible with the classical form of Darwinian genetic inheritance. It is also possible that such 'escapee' genes could result in brain disorders should they fail to achieve demethylation and thus remain silent during these later stages of development when they are needed for the brain.

In summary, not only is there a bias for epigenetic influences on gene expression through the matriline (see also Chapters 3 and 5), but from the very earliest stages of germ cell (sperm and egg) production, it is the matriline which plays a leading role in the reprogramming process of these germ cells. The inheritance of maternal mitochondrial DNA, certain DNA repair mechanisms and post-fertilisation development of trophectoderm for the next generation's placenta are all dependent on the selective expression of maternal alleles. It is becoming clear that the selection pressures, introduced by the viviparous in-utero development of mammals, have endowed the matriline with a biased contribution to the molecular epigenetic processes that ensure her reproductive success and foetal survival (Bourc'his & Bestor, 2006).

THE EPIGENETIC CLOCK

During embryo development, metabolism is one of the many cellular processes that is under continuous change, and sirtuins have emerged as one of the important regulators of cellular metabolism, the cell cycle, cellular apoptosis and autophagy. The deacetylase activity of histones depends on sirtuins, a function that is closely linked to cellular energy consumption. Moreover, this activity of sirtuins ultimately links the regulation of chromatin dynamics to glucose homeostasis, ageing and cellular lifespan. Sirtuins are histone deacetylases, thereby regulating gene expression and chromatin function in response to cellular metabolism (Rajendran *et al.*, 2011) (Figure 2.2).

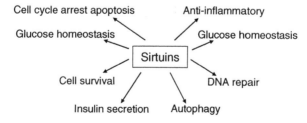

FIGURE 2.2 Regulation of cellular functions by sirtuins. Sirtuins are important regulators of many aspects of cellular function (metabolism, cell cycle, apoptosis, and autophagy) by regulation of chromatin structure.

Genetic studies of longevity and the search for genes that might contribute to a longer life have met with little success. In recent years, the studies of longevity have focussed more on the epigenetic processes that might provide a basis for a longer life. It is certainly the case that different individuals show considerable epigenetic variability with ageing. This is reflected across different tissues, and even found to vary for different cells within the same tissue. For the most part these cellular epigenetic patterns of gene expression are locked into each tissue type and are transmitted via mitosis into their developing daughter cells. It is this heritable epigenetic propagation of cells which potentially provides for more variance than is provided by somatic DNA mutations. Much of this cellular variance is brought about by metabolic factors which provide for an imperfectly maintained cellular milieu, thereby enabling the accumulation of epigenetic drift over time. Monozygotic twin studies have consistently revealed epigenetic-induced expression changes to individual genes not only during their development, but also with ageing. However, cells do tend to apply restraints on 'epigenetic drift' via methylation at CpG islands. This results in a lower level of epigenetic variation, but even this is not as stringent as that which occurs via the direct silencing of DNA coding regions. What, therefore, might be the advantages accruing from epigenetic variability? Considered from an evolutionary perspective, epigenetic variability may carry forward certain advantages by increasing the fitness and survival potential of individuals in ever-changing environments.

Cells are able to sense changing environments and different cells become specialised to different environments, translating these differences into specific modulations of the genome via chromatin remodelling. Indeed, links have been found between cellular metabolism and epigenetic control mechanisms (Berger & Sassone-Corsi, 2016). Metabolism is now thought to provide the links for the ageing of the senescent epigenome by changes to histone methylation and acetylation. For example, p53 deacetylation by sirtuins (silent information regulators) has been linked to cellular senescence. In laboratory animals, nuclear Sirt6 has been shown to have positive links with ageing by maintaining chromatin integrity (Masri *et al.*, 2014). Deletion of the *Sirt6* gene shortens the lifespan of the laboratory mouse, a condition brought about due to impairment of the base excision DNA repair pathway.

Restrictions on calorie intake have also been shown to have a beneficial effect on extending the lifespan of an organism via the rapamycin signalling pathway (Kaeberlein *et al.*, 2005), and again via the activation of sirtuins (Cohen *et al.*, 2004). Thus, evidence is now accumulating to support the theory that senescence and premature ageing are characterised and determined by restructuring the epigenome. The mechanisms that transduce metabolic signals into epigenetic changes involve the DNA deacetylase and acetylase enzymes. These in turn respond to changes in the cellular milieu of metabolites acting as co-factors that change epigenetic modulator enzymes. Senescence also induces an irreversible cell-cycle arrest in response to oncogenetic stress, which may provide a means for protecting the ageing body against cancer. Moreover, the removal of senescent cells is thought to ameliorate the effects of age-associated pathogens. However, which changes to chromatin represent those age-associated degradative changes, and which changes provide for protection from ageing, have yet to be clarified.

EPIGENETICS, BRAIN DEVELOPMENT AND FUNCTION

The brain is a truly exceptional biological structure. First and foremost it has an exceedingly long period of development, especially in humans, being the last organ in the body to reach maturity in the years following puberty. During this late-adolescent phase of brain development, a global reorganisation of the neocortex takes place and continues in humans until the early twenties. The neocortex is made up of six layers of cells, and this six-layered sheet has expanded substantially during mammalian evolution to provide a folded concertina-like structure (sulci and gyri) that occupies most of the skull's cranial cavity. Folding of the cortical sheet (gyrification) is a space-saving feature, providing more functional units (cortical columns), the number of which reaches the pinnacle of evolutionary progression in the human brain. The phylogenetically older parts of the brain become enclosed by this folded cortical sheet, and these so-called areas of 'limbic brain' regulate emotion and those primary-motivated behaviours which are common to all mammals. These limbic neural functions embrace feeding, sexual behaviour, fear, and aggressive behaviour in males and maternal care in females. Maintenance of sleep/awakening cycles and regulation of body temperature are also a functional feature of the limbic brain. Such functional attributes are essential for mammalian survival, and underpin vital physiological needs as well as regulation of emotional behaviour. The larger 'executive' neocortex is integral to learning and memory, a function that is accomplished by changes in synaptic connection strengths among its neural networks. Strengthening of neural interconnections engages a genetic programme that is common to all neurons, but which of these connections are strengthened is notably dependent on their activity. Such activity is generated by the environment and interpersonal experiences, and hence the consolidation of neural connection strengths is linked to the epigenome.

These distinct regions of the brain (executive and emotional) are not functionally independent. The executive neocortex, especially in humans, has a strong regulatory influence over the phylogenetically ancient 'emotional brain' which it encloses. It is the regions of the emotional brain that interact very purposefully with the hormonal messengers from the body in the context of hunger, satiety, fear, aggression and sex. The neocortex performs more rational decision making based on experience and, in humans, the social rules and laws drawn up by the experiences of previous generations. Only then does the neocortex decide how to react, or not react, depending on the context. The larger the executive neocortex of mammals, the less deterministic are the hormones, and the more influential is a rationalisation for the behavioural actions to be undertaken. In this way, the mammalian brain can be conceptualised as the organ that co-adaptively integrates the internal world of the body (hormonal messengers) with the external environmental world, most notably that of the social world in humans. Brain evolution has seen an emancipation from the hormonal genetic regulation of behaviour, but a significant gain for the influence of social factors in the regulation of behavioural interactions. Not only are these social interactions important for regulating behaviour, but they also play a significant role in regulating the hormone secretions which now serve as a 'back-up'.

Interestingly, the mammalian female's brain, when compared with that of the male brain, has an additional set of hormonal genetic modifiers that need to be taken into account, namely those hormones from the placenta when the female is pregnant (Keverne, 2006). We tend to think of the placenta as belonging to the mother and often refer to it as the maternal placenta. This is genetically incorrect. The placenta is foetal, produced by genes originating from both father and mother. Hence it is the foetus, via its placental hormones, which hormonally regulate the mammalian mother's physiology and behaviour (Figure 2.3). The primary motivated behaviours of maternal, sexual and feeding behaviour are determined by the mother's hypothalamus, which together with the amygdala influence motivation and emotion. These functions are programmed by the action of hormones from the next generation's foetal placenta. In this way the mother's sexual behaviour is curtailed, and feeding behaviour is enhanced during the period of pregnancy in most mammals. The foetal hormones from the placenta further regulate the mother's provision of milk, and activate maternal nest building, as well as priming the mother's brain for maternal care. Hence, early in foetal development, the hormones of the foetal placenta are acting on the mother's brain, instructing her to eat more in advance of the later demands that will be made by the growing foetus. Placental progesterone, in particular, increases maternal food intake early in pregnancy, way ahead of the subsequent foetal energetic requirements that will arise in the later stages of pregnancy. Indeed, it would not be possible for

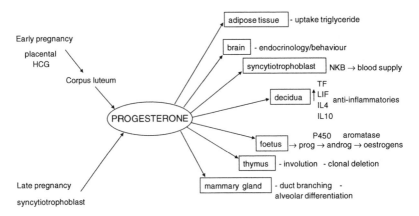

FIGURE 2.3 Placental progesterone production and maternal target tissues. The foetus via its placental hormones transgenerationally regulates many aspects of the mother's physiology and behaviour via several target tissues. Progesterone is particularly important throughout pregnancy, especially for actions on the brain.

most mothers to find and consume sufficient amounts of food at this later stage of pregnancy. Hence the early laying down of these maternal energy stores is determined by the genome of the foetal placenta. This is transgenerationally relayed to the mother by the action of the foetal placental hormones. The success of this intergenerational hormonal transmission is essential for the success of foetal survival in the later stages of pregnancy. These same foetal–placental hormones also activate the mother's brain to synthesise the hormone oxytocin in anticipation of the requirement for this hormone at parturition for activation of the birth process, and for subsequent regulation of maternal behaviour and milk ejection. A further foetal–placental hormone (allopregnenalone) has protective properties for neurons against ischaemia and the lack of oxygen in both the mother's brain and in the foetal brain. This same hormone restrains maternal stress and prevents the release of stress hormones that can promote premature birth (Knight et al., 2012). In short, the foetal–placental genome determines its own destiny by epigenetically and intergenerationally regulating the mother's 'emotional' brain. This serves the interests of the foetus which, at the same time, is in the process of developing these very same hypothalamic regions within its own emotional brain. Mammalian in-utero development has thus provided a new dimension on which evolutionary selection pressures may operate, namely that of the intergenerational action of hormones that determine the expression of those genes which are maternally epigenetically imprinted for mono-allelic gene expression. The mother's uterine environment has thus provided strong heritable effects, especially on the female foetus. This developmental legacy is transferred to daughters but not to sons due to the early male development of foetal Leydig cells which masculinise the male's brain by the production of testosterone.

These findings raise the question as to how the developing placental genome belonging to the next generation 'knows' the optimal way of regulating the adult maternal 'emotional' brain of the present parental generation. Likewise, how has the adult maternal brain developed to respond correctly to the future demands of the next generation (Keverne, 2015)? These intergenerational co-adaptive events also require co-adaptations across the maternal and foetal genomes, the success of which is epigenetically carried forward to the next generation through genomic imprinting (see Chapters 3 and 5 for more detail). The same epigenetically imprinted genes which develop both the foetal 'emotional brain' and foetal placenta do so at a time when the placenta is instructing the previous maternal generation's 'emotional' brain for provisioning of the foetus. In this way, the developing 'emotional' brain and developing placenta serve as a unitary template on which positive selection pressures for good mothering operate across successive generations (Keverne, 2014b). Such co-adaptive transgenerational events are driven by the effectiveness and the success of the foetal placenta's interactions with the maternal 'emotional' brain of the previous generation. Thus, offspring which receive both optimal nourishment and attentive maternal care will themselves develop an 'emotional' brain that is both genetically determined and epigenetically predisposed to good mothering for the next generation. The adult male brain never experiences the presence of a placenta or the effect of placental hormones and is therefore not evolutionarily predisposed to be parental in the same way.

Yet another linkage between the placenta and mother's brain is the finding that placental stem cells from the human full-term placenta can be epigenetically induced to differentiate into neurons. As far as I am aware, this does not take place during normal development, but it does illustrate commonalities across those epigenetically imprinted genes which regulate both the brain and placenta. Other studies have shown the placenta itself to be a source of the neurotransmitter serotonin. Whether this blood-borne serotonin can influence the development of the foetal brain is not yet known. However, it is known that placental serotonin can increase the kind of placental enzyme activity which does have a controlling action on those placental calming hormones which in turn have important influences on the mother's 'emotional' brain during pregnancy.

NEOCORTICAL DEVELOPMENT, EPIGENETICS AND SOCIAL INFLUENCES

The brain is a very specialised organ endowed with the unique ability to respond to its environment, reshaping its connections' strengths according to what the brain's neurons have experienced (Keverne, 2014a). How the brain develops is thus influenced not only by the relatively stable in-utero environment, but more importantly by its extended period of postnatal growth and development, which takes place in an ever-expanding social environment. This maturation of the human brain endures adolescence and lasts until the early twenties in humans. Such a very long maturational phase for development of the human brain takes place in the less-secure, indeed sometimes destabilising and certainly ever-changing social environment. Because the making and consolidation of the brain's interconnections are dependent on the activity of its neurons, then it stands to reason that the kind of environment creating this activity has a substantial impact on which connections are made, and on the consolidation of their synaptic strengths. As already mentioned, neural connection changes are brought about according to the epigenetic regulation of gene transcription. The very nature of epigenetic regulatory control over gene transcription means that different neurons, all possessing the same genome, can be developmentally regulated in different ways using the same overlapping gene networks. The brain is thus very special. It has a life history of its own that is environmentally driven according to the information it receives, and thus is evolutionarily different from the rules that govern most of the body's other organs.

Certain genes that are destined to form the neocortex have a late phase of reprogramming in the postnatal period (Tuesta & Zhang, 2014). Indeed, the major part of neocortical brain development occurs postnatally, and its functioning is specialised to adapt to the environment, both physical and social. Not unlike germline cells, cortical neurones also require a capacity to reprogramme throughout adult life by demethylation/remethylation. This occurs during the process of learning and memory (Kaas et al., 2013) and during adult neurogenesis which particularly impacts on the hippocampus and the olfactory system. Learning and memory require neuronal activity which can strengthen active synaptic connection strengths and weaken other inhibitory synaptic strengths through a process of synaptic plasticity. The signalling events triggered by synaptic activity involve calcium entry to the neuron via a range of neural receptors, a process that underpins learning and memory (NMDA and GABA receptors feature predominantly). This leads to changes in neurotransmitter release as well as epigenetic changes to nuclear DNA methylation, histone acetylation and hydroxyl methylation. As a result of these epigenetic modifications to nuclear DNA, neurons gain a high state of plasticity to integrate and store new information. Inhibition of Dnmt (the DNA methyl transferases) interrupts long-term potentiation in the hippocampus and changes levels of methylation in the promoter regions of *Reelin* and *BDNF* genes. These are two of the hippocampal genes engaged with, and essential for, synaptic plasticity (Levenson et al., 2006).

Tet1,2,3 methylcytosine dioxygenases play an integral role in the programming of the embryonic germline, but in neurons these same enzymes are equally crucial to the mechanisms underpinning learning and memory (Kohli & Zhang, 2013). Indeed, Tet-mediated demethylation is found at its highest levels in the brain, and is thought to influence passive demethylation by acting on the *Dnmt* enzymes, and thereby serve as an intermediary in active demethylation. Tet1 and Tet2 methylcytosine dioxygenases have been shown to promote DNA demethylation in the adult brain at activity-dependent synapses in the hippocampus (Kaas et al., 2013). Analysis of *Tet1* gene deletion in mice has been shown to downregulate hippocampal genes that are engaged with neural activity, and this is accompanied by impaired memory extinction and abnormal LTP (long-term depression) in the hippocampus (Rudenko et al., 2013). Mice with gene mutations to *Tet1* have also been shown to have impaired short-term learning and memory (Zhang et al., 2013). Moreover, these *Tet1*-depleted mice show no developmental abnormalities in the brain, or in neuronal neurogenesis, suggesting a level of redundancy across the methylcytosine dioxygenases.

Certain regions of the brain (hippocampus and olfactory bulbs) continue to receive developing migratory neurons even in the adult brain. Overexpression of *Tet3* has been shown to disrupt the successful integration of these migratory inhibitory GABA-ergic neurons from the brain's subventricular zone on their way to the olfactory bulb (Colquitt et al., 2013). The continuous renewal of these inhibitory GABA-ergic neurons is required to expand neural plasticity in the olfactory system, which is the most important sensory system for the majority of small-brained mammals. Moreover, the continual exposure of the olfactory receptor neurons to environmental toxins results in their premature cell death. Hence, both their replacement and regeneration throughout life is also essential to the continuity of chemosensory functioning. Such regeneration of these receptor neurons further requires a degree of reorganisation when their axons reach their tri-lamina olfactory bulb circuitry. *Tet3* action is therefore important to ensure the functional reintegration of the axons from these migratory inter-neurons.

INTERGENERATIONAL INHERITANCE

Intergenerational inheritance, especially in humans, is primarily achieved through the way in which the postnatal social environment shapes the epigenetic regulation of neural connection strengths, especially of those neurons in the developing neocortex of this next generation (Szyf, 2013). Economically privileged families tend to produce economically privileged children; academic families tend to produce academic children; aggressive families tend to produce aggressive children; obese families tend to produce obese children; and abused children tend to show abuse in the next generation. These events only provide for a 'tendency', and such underpinning brain development is certainly not genetically predetermined. However, unless this cycle of socially dependent events is disrupted, the brain becomes epigenetically adapted to the kind of social environment in which it has developed (Champagne, 2012). This is particularly true of the neocortex, which develops into the largest component of the human brain. However, for the majority of individuals, much of this neocortical developmental time is usually spent under the protective environment of the caring extended family. The importance of this social environment cannot be emphasised enough, and it is the mother, in particular, who becomes the significant attachment figure and provides the secure base from which other relationships develop and social horizons broaden (Keverne, 2004).

This dependence on adaptive epigenetic changes for normal development of the brain's connections may also produce dysfunction if the environment is chronically stressful. However, most of these stressful experiences are transient and readily overcome. Other experiences may be more durable, especially those which are chronically aversive, occur for long periods when the brain is developing, and thereby subject the brain to epigenetic modification. Animal studies have shown a causal relationship between early life adversity and changes in epigenetic methylation to DNA which occurs for a substantial number of genes expressed in the brain. Methylation silencing of genes is one epigenetic mechanism which, in rhesus monkeys, is influenced in the prefrontal cortex by rearing conditions. Rearing monkey offspring with a doll-like artificial surrogate mother results in their adult maternal caring being inadequate, and is accompanied by different patterns of gene methylation in their cerebral cortex (Provencal *et al.*, 2012). Other studies have shown distinct histone methylation signatures in the frontal cortex of schizophrenic brains, and DNA methylation differences assessed from the blood in genetically identical twins discordant for schizophrenia. Interpreting methylation findings from blood samples may not reflect what occurs in the brain, but as human brain samples are problematic to obtain, they certainly provide a cellular indicator for further consideration.

The neocortex is the largest part of the human brain and those regions (frontal and temporal cortex) which are uniquely enlarged in human species are the last to develop and refine their interconnections in the post-pubertal period until around 22 years of age in humans (see Chapter 4). Thus, the early maternal environment progressively becomes one small part of an ever-expanding social world as age progresses. In contrast to the in-utero environment, this world becomes progressively more variable and risky. Moreover, the complexity of the interconnections made by the neocortex and the millions of neurons and billions of synapses that are made inevitably means that developmental errors are inevitable. However, those neurons which fail to make the right connections at the right time undergo programmed cell death. Of course, there is no hard-wired genetic programme which ordains when and where cell death occurs; this again is dependent on neural activity. Moreover, all cortical neurons make contact with other neurons that undertake the same process. Because consolidation of neural connections is activity-dependent, then the environment that generates this activity is also clearly important. Thus, no two brains are exactly the same, even in genetically identical twins, and although genetic programmes clearly play an integral part in brain development, so too does the environment via the regulation of epigenetic gene expression. This ensures that not only the socially positive, but also those chronically adverse conditions are 'imprinted' across neural DNA transcription in determining connectivity strengths.

Because the kind of environment which is generating neural activity determines the trajectory for brain functional development, then it is no surprise that infants who are reared in extremely deprived environments fail to develop language and continue to crawl rather than to walk (Levin *et al.*, 2014). With interventions and subsequent persistent attention, these children usually do learn the ability to walk, although their mobility is considerably delayed compared with that of the normal toddler. Abnormal neural connections in the brain are also found in children that experience severe socioemotional deprivation. Neurophysiological assessment of such infants has revealed cognitive impairments and impulsive behaviour, which is reflected in structural changes of the brain's cortical connections, especially between the temporal and frontal neocortex (Fox *et al.*, 2011). Other studies have found abnormalities in language, and in those emotionally driven interconnecting pathways of the brain. This has been monitored even at 10 years of age, especially in those children who have experienced histories of early social deprivation (Kumar *et al.*, 2014). Most of life's adverse experiences are transient and forgotten. However, experiences which are traumatic and never forgotten do epigenetically predispose the brain's circuitry to general emotional arousal, even in contexts that are only mildly reminiscent. Some of these contexts may even be generated by the imaginary thought processes of the brain itself.

The advantages gained from epigenetic variability are very important for human cortical brain development, but the underpinning neural mechanisms are equally capable of providing a susceptibility for adverse functional outcomes. It would thus appear that artistic creativity is often linked with extremes in mood, thought and behaviour, which may sometimes include bipolar depression. It is also the case that the characteristics of 'mania' and 'genius' likewise share some similar developmental brain origins. Psychopathological symptoms, especially those belonging to bipolar mood disorder, and manic depression occur more frequently in poets, writers, composers and artists (Jamison, 1989). Compared with the general population, bipolar mood disorder is significantly over-represented among creative artists and this creativity is often punctuated by periods of severe depression. Many creative individuals experience far from satisfactory social interactions and family lives. Perhaps more of a concern is the outcome from manic depression, which can also become a force for antisocial behaviour and even psychopathology. In the extreme cases of Napoleon, Hitler and Stalin, manic depression drove them to seek absolute power and abuse of this power resulting in mass killing, war on their perceived enemies, and even on their own countrymen (Leib, 2008).

In conclusion, the human brain is a truly exceptional organ, first and foremost in its exceedingly long period of development. The brain's capacity to renew and to change its connection strengths with other neurons, especially in the context of environmental experiences, is integral to accomplishing learning and memory. The brain also needs to forget and to consolidate new information by revision of established memories. Like the germline cells of the ovary and the testis, the neurons of the brain have the capacity for genome reprogramming, but in the case of neurons this is not limited to a very discrete, early period of development. Mechanistically, learning requires neuronal activity to strengthen the brain's interconnections and synapses, and to weaken other connection strengths by a process of synaptic plasticity. Epigenetically, this leads to changes in DNA histone methylation in neurons. Forgetting or changing these learning experiences employs the same DNA demethylation enzymes that are active in reprogramming the germline, the so-called *TET* enzymes. This makes the brain very special. It has a life history of its own that is epigenetically dependent but, in part, evolutionarily independent of the rules that govern other body tissues. The brain transmits information to other brains intergenerationally and transgenerationally, even across multiple generations through the written word and social media. Brains on the receiving end can choose to assimilate or to ignore this information. Not all individuals are equal in the eyes of others, and hence some social interactions have more impact than others. In this way, leadership is acquired, new standards are set and new laws are created, both promoting and providing constraints on the working of the brain in this and in future generations. New cultures develop, extending the social parameters for learning, and extending cross-cultural knowledge around the world. Having transcended the slow and environmentally constrained progress of biological evolution, the brain can advance human social evolution at relatively high speed. In recent generations this has been especially enhanced by the intervention of electronic brains (computers) that make communication even faster, multidimensional and international.

Since the early theories of mother–infant attachment and bonding were formulated, developmental science has provided insights as to how lifetime interactions have shaped neurobiological outcomes. Following the sequencing of the human genome, an intense interest developed as to how genetics may have contributed to this understanding of development. Single nucleotide polymorphisms (SNPs, pronounced 'snips') and single-point mutations to gene structure became a focus for attention during the past decades. Unfortunately this approach has revealed very little specific information, especially the much-needed information required for understanding brain dysfunction. An evolutionary comparison of brain genetics has also revealed that very few differences in gene coding sequence exist. Even between the brain of chimpanzees and humans there are few differences in protein coding genes, and yet the human brain is three times the size of this primate ancestor. Moreover, the majority of total human gene transcripts (some 76 per cent) are expressed in the brain, but have a variety of mechanisms available for differential regulation. This would all point to changes in gene regulatory mechanisms, the 'where' and 'how' and 'when' in the brain these regulatory mechanisms are called upon, as being more informative for the maturational development of the brain, than any deterministic action of genes.

There is certainly potential for error, but surprisingly development mainly gets it right, albeit introducing considerable variability across individuals. It is this variability which particularly characterises the human brain – no two brains are identical, even when the genomes are identical in monozygotic twins. Brains are extremely adaptable. Such adaptability carries with it a multiplicity of advantages, but there is also a high risk for problems if we fail to constrain the brain's adaptability along a pathway that secures its well-being.

REFERENCES

Berger, S. L. & Sassone-Corsi, P. (2016). Metabolic signaling to chromatin. *Cold Srping Harb. Perspect. Biol.* 8: a019483.

Bhan, A. & Mandal, S. S. (2016). Estradiol-induced transcriptional regulation of long non-coding RNA, HOTAIR. *Methods Bol. Biol.* 1366: 395–412.

Biason-Lauber, A. & Chabolissier, M. C. (2015). Ovarian development and disease: the known and the unexpected. *Semin. Cell Dev. Biol.* 45: 59–67.

Bird, A. (2007). Perceptions of epigenetics. *Nature* 447: 396–98.

Bourc'his, D. & Bestor, T. H. (2006). Origins of extreme sexual dimorphism in genomic imprinting. *Cytogenet. Genome Res.* 113: 36–40.

Champagne, F. A. (2012). Interplay between social experiences and the genome: epigenetic consequences for behavior. *Adv. Genet.* 77: 33–57.

Cohen, H.Y., Miller, C.A., Bitterman, K. J., et al. (2004). Calorie restriction promotes mammalian cell survival by inducing the SIRT1 deacetylase. *Science* 305: 390–92.

Colquitt, B. M., Allen, W. E., Barnea, G., et al. (2013). Alteration of genic 5-hydroxymethylcytosine patterning in olfactory neurons correlates with changes in gene expression and cell identity. *Proc. Natl Acad. Sci. USA* 110: 14682–87.

Comings, D. E. (1972). The genetic organisation of chromosomes. *Adv. Hum. Genet.* 3: 237–431.

Coufal, N. G., Garcia-Perez, J. L., Peng, G. E., et al. (2011). Ataxia telangiectasia mutated (ATM) modulates long interspersed element-1 (L1) retrotranposition in human neural stem cells. *Proc. Natl Acad. Sci. USA* 108: 20382–87.

Daxinger, L. & Whitelaw, E. (2012). Understanding trangenerational epigenetic inheritance via the gametes in mammals. *Nat. Rev. Genet.* 13: 153–62.

de Koning, A. P., Gu, W., Castoe, A., et al. (2011). Repetitive elements may compromise over two-thirds of the human genome. *PLoS Genet.* 7: e1002384.

El Najj, N., Schneider, E., Lehnen, H., et al. (2014). Epigenetics and life-long consequences of an adverse nutritional and diabetic intrauterine environment. *Reproduction* 148: R111–20.

Erwin, J. A., Marchetto, M. C. & Gage, F. H. (2014). Mobile DNA elements in the generation of diversity and complexity in the brain. *Nat. Rev. Neurosci.* 15: 497–506.

Fox, N. A., Almas, A. N., Degnan, K. A., et al. (2011). The effects of severe psychosocial deprivation and foster care intervention on cognitive development at 8 years of age: findings from the Bucharest Early Intervention Project. *J. Child Psychol. Psychiatry* 52: 919–28.

Gall, J. G. (1981). Chromosome structure and the C-value paradox. *J. Cell Biol.* 91: 3s–14s.

Gurdon, J. B. (2013). The egg and the nucleus: a battle for supremacy. *Development* 140: 2449–56.

Hatanake, Y., Shimizu, N., Nishikawa, S., et al. (2013). GSE is a maternal factor involved in active DNA demethylation in zygotes. *PLoS ONE* 8: e60205.

Heard, E. & Martienssen, R. A. (2014). Transgenerational epigenetic inheritence: myths and mechanisms. *Cell* 157: 95–109.

Jamison, K. R. (1989). Mood disorders and patterns of creativity in British writers and artists. *Psychiatry* 52: 125–34.

Jedrusik, A., Bruce, A. W. & Tan, M. H. (2010). Maternally and zygotically provided *Cdx2* have novel and critical roles for early development of the mouse embryo. *Dev. Biol.* 344: 66–78.

Jenkins, T. G. & Carrell, D. T. (2012). Dynamic alterations in the paternal epigenetic landscape following fertlization. *Front. Genet.* 3: 143.

Kaas, G. A., Zhong, C., Eason, D. E., et al. (2013). *TET1* controls CNS 5-methylcytosine hydroxylation, active DNA demethylation, gene transcription, and memory formation. *Neuron* 79: 1086–93.

Kaeberlein, M., Powers, R. W. 3rd., Steffen, K. K., et al. (2005). Regulation of yeast replicative life span by *TOR* and *Sch9* in response to nutrients. *Science* 310: 1193–96.

Keverne, E. B. (2004). Understanding well-being in the evolutionary context of brain development *Phil. Trans. R. Soc. Lond. B* 359: 1349–58.

(2006). Trophoblast regulation of maternal endocrine function and behaviour. In: A. Moffett, C. Loke & A. McLaren (eds.), *Biology and Pathology of Trophoblast*. New York, NY: Cambridge University Press, pp. 148–63.

(2014a). Significance of epigenetics for understanding brain development, brain evolution and behaviour. *Neuroscience* 264: 207–17.

(2014b). Mammalian viviparity: a complex niche in the evolution of genomic imprinting. *Heredity* 113: 138–44.

(2015). Genomic imprinting, action, and interaction of maternal and fetal genomes. *Proc. Natl Acad. Sci. USA* 112: 6834–40.

Knight, S. R., Davidson, C., Young, A. M., et al. (2012). Allopregnanolone protects against dopamine-induced striatal damage after in vitro ischaemia via interaction at GABA A receptors. *J. Neuroendocrinol.* 24: 1135–43.

Kohli, R. M. & Zhang, Y. (2013). TET enzymes, TDG and the dynamics of DNA demethylation. *Nature* 502: 472–79.

Kumar, A., Behen, M. E., Singsoonsud, P., et al. (2014). Microstructural abnormalities in language and limbic pathways in orphanage-reared children: a diffusion tensor imagining study. *J. Child Neurol.* 29: 318–25.

Kung, J. T. Y., Colognori, D. & Lee, J. T. (2013). Long noncoding RNAs: past, present, and future. *Genetics* 193: 651–69.

Leib, J. (2008). Two manic-depressives, two tyrants, two world wars. *Med. Hypotheses* 70: 888–92.

Levenson, J. M., Roth, T. L., Lubin, F. D., et al. (2006). Evidence that DNA (cytosine-5) methyltransferase regulates synaptic plasticity in the hippocampus. *J. Biol. Chem.* 281: 15763–73.

Levin, A. R., Seanah, C. H., Fox, N. A., et al. (2014). Motor outcomes in children exposed to early psychosocial deprivation. *J. Pediatr.* 164: 123–29.

Lewin, B. (2006). *Essential Genes*. Upper Saddle River, NJ: Pearson Education Inc.

Li, X., Ito, M., Zhou, F., et al. (2008). A maternal-zygotic effect gene, *Zfp57*, maintains both maternal and paternal imprints. *Dev. Cell* 15: 547–57.

Malki, S., van der Heijden, G. W., O'Donnell, K. A., et al. (2014). A role for retrotransposon LINE-1 in fetal oocyte attrition in mice. *Dev. Cell* 29: 521–33.

Marchetti, F., Essers, J., Kanaar, R., et al. (2007). Disruption of maternal DNA repair increases sperm-derived chromosomal aberrations. *Proc. Natl Acad. Sci. USA* 104: 17725–29.

Masri, S., Rigor, P., Cervantes, M., et al. (2014). Partitioning circadian transcription by *SIRT6* leads to segregated control of cellular metabolism. *Cell* 158: 659–72.

Nakagawa, M. & Yamanaka, S. (2010). Reprogramming of somatic cells to pluripotency. *Adv. Exp. Mol. Biol.* 695: 215–24.

Ohno, S. (1972). So much 'junk' DNA in our genome. In: H. H. Smith (ed.), *Evolution of Genetic Systems*. New York, NY: Gordon & Breach, pp. 366–70.

Provencal, N. M., Suderman, M. J., Guillemin, C., et al. (2012). The signature of maternal rearing in the methylome in rhesus macaque prefrontal cortex and T cells. *J. Neurosci.* 32: 15626–42.

Rajendran, R., Garva, R., Drstic-Demonacos, M., *et al.* (2011). Sirtuins: molecular traffic lights in the crossroad of oxidative stress, chromatin remodeling, and transcription. *J. Biomed. Biotechnol,* 2011: 368972.

Reitz, C., Tosto, G., Mayteaux, R., *et al.* (2012). Genetic variants in the fat and obesity associated (*FTO*) gene and risk of Alzheimer's disease. *PLoS ONE* 7: e50354.

Rudenko, A., Dawlaty, M. M., Seo, J., *et al.* (2013). Tet1 is critical for neuronal activity-regulated gene expression and memory extinction. *Neuron* 79: 1109–22.

Surani, M. A., Hayashi, K. & Hajkova, P. (2007). Genetic and epigenetic regulators of pluripotency. *Cell* 128: 747–62.

Szyf, M. (2013). How do environments talk to genes? *Nat. Neurosci.* 16: 2–4.

Tang, W. W., Kobayashi, T., Irie, N., *et al.* (2016). Specification and epigenetic programming of the human germ line. *Nat. Rev. Genet.* 17: 585–600.

Tuesta, L. M. & Zhang, Y. (2014). Mechanisms of epigenetic memory and addiction. *EMBO J.* 33: 1091–103.

Turelli, P. (2014). Interplay of TRIM and DNA methylation in controlling human endogenous retroelements. *Genome Res.* 24: 1260–70.

Waddington, C. (1946). *How Animals Develop.* London: George Allen & Unwin Ltd.

Wossido, M., Arand, J., Sebastiano, V., *et al.* (2010). Dynamic link of DNA demethylation, DNA strand breaks and repair in mouse zygotes. *EMBO J.* 29: 1877–88.

Wu, J., Huang, B., Yin, Q., *et al.* (2016). The landscape of accessible chromatin in mammalian preimplantation embryos. *Nature* 534: 652–57.

Zhang, R. R., Cui, Q. Y., Murai, K., *et al.* (2013). Tet1 regulates adult hippocampal neurogenesis and cognition. *Cell Stem Cell* 13: 237–45.

Zhao, J., Sun, B. K., Erwin, J. A., *et al.* (2008). Polycomb proteins targeted by a short repeat RNA to the mouse X chromosome. *Science* 322: 750–56.

3 Genomic Imprinting: Matrilineal Regulatory Control Over Gene Expression

The evolution from egg-laying monotreme mammals to the placental marsupial mammals not only witnessed the arrival of a Y-chromosome and the so-called male sex-determining gene, but with it came a new mechanism for the control of autosomal gene regulatory expression, namely 'genomic imprinting' (Renfree et al., 2013). This epigenetic regulatory mechanism results in certain autosomal non-sex-determining genes becoming expressed according to their chromosomal parent of origin. In other words, some of our autosomal genes are only expressed when we inherit them from our mothers and others are only expressed when inherited from our fathers. Autosomal genes are constructed on two strands of DNA, a maternal copy and a paternal copy, from which the genetic alleles, maternal and paternal, are usually both expressed. This more usual biallelic gene expression is translated into RNA which provides the genetic code for assembling the manufacture of amino acids. From these amino acids, construction of the protein building blocks of the cell are made, which in turn produce the tissues of the body. An equivalent of autosomal gene expression according to parent of origin 'imprinting' is present in some insects, but among vertebrates it appears to be unique to those mammalian species with a placenta. Genomic imprinting has abandoned gene expression from both alleles in favour of consistently expressing only one allelic copy of the gene, either maternal or paternal, but always expressing this same allelic copy in the cells of both sons and daughters. In this way, regardless of sex, some imprinted autosomal genes are only expressed when they originate from the mother, whereas others are expressed only when they originate from the father. The most prevalent mechanism for determining genomic imprinting is brought about by an 'epigenetic' chromosomal mark, the imprint control region or ICR. This methylation mark on DNA is heritable through the matriline, thereby providing the mother's genome with control over which imprinted allele is expressed, maternal or paternal (Bourc'his & Bestor, 2006). Paternal expression occurs when the ICR is of maternal origin, but when this ICR regulates inhibitory non-coding micro-RNAs embedded in the imprinted gene cluster, then maternal expression follows. Throughout mammalian evolution, more autosomal genes have been recruited to the ICR, thereby resulting in their imprinted expression. Moreover, a number of imprinted gene clusters also contain small non-coding RNAs which are regulated by the Drosha/Dgcr8 complex. These non-coding RNAs are gaining recognition in the context of brain development and psychiatric disorders, as well as in placental development and placental dysfunctions, regions for which imprinted genes are heavily engaged.

HISTORICAL PERSPECTIVE ON GENOMIC IMPRINTING

The first indirect evidence for inheritance being governed by different 'parent of origin' autosomal genes – that is, genes not involving the XY sex chromosomes – arose from the study of hybrid mammalian crosses (Gray, 1972). These hybrids are produced by reciprocal crossmating of different mammalian species and results in offspring of the same genetic make-up, but which differ in appearance according to parental mating. Thus, when a male horse (stallion) is mated with a female donkey, a 'hinny' is produced, and conversely if a female horse (mare) is mated with a male donkey, a 'mule' is produced. Both reciprocal crosses show exceptional hardiness and endurance, but both are sterile and differ from each other in size and general appearance, even though they each possess the same genomes. It was quite fashionable towards the middle of the last century to produce exotic hybrids (e.g. crosses between lions and tigers) which were produced by reciprocal cross-breeding of these carnivores in captivity. When the male was a lion and the female was a tiger, more successful births were achieved (named ligers) than with the reciprocal cross from a male tiger mating with a female lion (named tigons). Tiger stripes in the fur are present in both types of reciprocal hybrid cross, as is the male lion mane, but the liger is much larger than either parental species or indeed the hybrid tigon. Again, back-crosses are extremely problematic to produce due to hybrid infertility. With our current knowledge of genomic imprinting, hindsight now informs us that imprinted gene regulation was responsible for these hybrids which, theoretically, were genetically identical but exhibited different phenotypes according to autosomal parental gene origin. Human hybrids, although rare, do exist and are usually found among female twins, where one individual from these offspring is made up from a mixture of cells from both sisters. Each subpopulation of cells maintains its own genetic identity to the extent that a cellular DNA test may miss-identify this sister as not belonging to the parents.

Genomic imprinting is exceptionally complicated and its understanding has required a phase-shift in our way of thinking about Mendelian inheritance. These new genetic insights arose from the studies of a reproductive biologist, Azim Surani, who was endeavouring to understand if parthenogenetic embryos (embryos containing two copies of maternal genes) could be created by the fusion of two female oocytes (unfertilised eggs) (Surani et al., 1984). Professor Surani worked in the laboratory of Bob Edwards, who was awarded a Nobel Prize for his work on 'in-vitro' fertilisation and the so-called 'test tube babies'. Surani employed these in-vitro techniques in his experiments with mice, and discovered the phenomenon of genomic imprinting when trying to understand the genetic basis for failure of 'parthenogenesis'. Parthenogenetic embryos, made from the fusion of two female egg nuclei (i.e. XX offspring born with two copies of the maternal genome, but no genes from the father), are also known as 'virgin births' because no male is involved in their conception. This procedure was purely experimental and is certainly not known to occur in nature. However, the 'test-tube baby' technology enabled this experimental work to be performed on mice, and involved the fusion of the nuclear germ cell DNA from two female mouse oocytes (parthenogenesis) in the absence of any male sperm DNA. These two copies of the maternal oocyte genome, when implanted into a mouse surrogate mother, were able to develop but only to the very earliest placental stages, and the process of foetal development was severely impaired. Similar studies were undertaken by removal of the nucleus from the female egg and replacing this with two fused copies of the male sperm nuclei (androgenetic embryos). This procedure also permitted embryonic development to reach a stage where implantation was possible, but again placental development was abnormal. However, this androgenetic placenta differed from that of the parthenogenetic placenta, in that two copies of the father's genes did produce a larger, albeit dysfunctional placenta (McGrath & Solter, 1984).

The intellectual excitement generated by these early experiments gave rise to a number of important questions. Namely, which genes are regulated by the parental imprinting mechanism; how are these genes expressed according to parent of origin; and finally, what function do these genes play in normal development? These questions were difficult to address at this stage because both parthenogenetic (Pg) and androgenetic (Ag) embryos fail to survive beyond the very earliest stages of placental development. In order to overcome such premature fatality, chimeric mouse embryos were developed. These embryos contained a mixture of normal cells (genetically disomic) together with cells that contained two copies of maternal genes (parthenogenetic chimeras) (Allen et al., 1995). Alternatively, embryos made up of normal cells together with cells containing two copies of paternal genes (androgenetic chimeras) were made for comparison. Genetically harmless identifier markers were also inserted into these uniparental male and female cells, enabling them to be traced and positionally mapped both during development and after birth (Figure 3.1). In both cases, embryo survival depended on no more than 50% of these uniparental chimeric cells (either Ag or Pg) constituting the embryo. With the important knowledge gained on the role these cells provided on placental and embryo growth differences

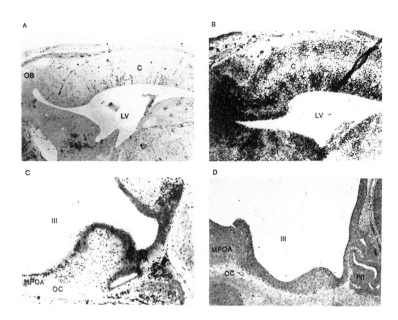

FIGURE 3.1 Distribution of Ag and Pg cells in chimeric mice. Distribution of parthenogenetic (Pg) cells, containing two copies of the maternal genome, proliferate in the executive neocortex (B; Pg cells shown in black), while androgenetic cells containing two copies of the paternal genome collect in the basal hypothalamus regions (C; Ag cells shown in black). Very few Ag cells are found in the cortical forebrain (A) and very few Pg cells thrive in the basal hypothalamic regions (D). OB, olfactory bulb; LV, lateral ventricle; C, cortex; III, third ventri; MPOA, medial preoptic area; OC, optic chiasm.

(early fatality induced by pure parthenogenetic or pure androgenetic embryos), the focus of these chimeric studies was to achieve successful embryo growth and survival in order to gain further knowledge on how the embryo itself was developmentally changed. Such studies were principally directed to providing a window on the brain, since the brain was known to be a target for the human 'parent of origin' genetic dysfunction seen in rare neural disorders (e.g. Prader–Willi syndrome). Using these chimeric cells in studies on mice, a clear and distinct patterning of distribution between the two types of cell (androgenetic and parthenogenetic) emerged in the developing brain

(Keverne et al., 1996). At birth, neural cells containing two copies of the paternal genome were found at very high levels in those parts of the limbic brain that are fundamentally common to all mammals (Figure 3.1, lower left). These are the evolutionary ancient areas of the 'emotional' brain, areas that are integral to the functioning of primary motivated behaviour and survival (feeding, sexual behaviour, reproductive fertility, maintaining body temperature, aggression), and these limbic brain regions were the target areas for the paternal androgenetic cells. Such genetically derived androgenetic cells were, however, excluded relative to control cells from those areas of the brain concerned with higher-order cortical brain functioning of the kind involved in the development of executive strategic planning. By contrast, parthenogenetic cells carrying two copies of the maternal genome were excluded from those brain areas (medio-basal hypothalamic forebrain) primarily concerned with motivated behaviour and survival. The parthenogenetic cells (exclusively female genome) did, however, selectively accumulate in the developing executive brain (neocortex and striatum) (Figure 3.1, upper right). Neocortical and striatal brain regions are actively concerned with planning for action, decision making and engaging the motor (movement) activity which carries forward these action plans. Furthermore, growth of these executive forebrain regions in parthenogenetic chimeras was enhanced by this increased female gene dosage brought about by the two copies of maternally expressed alleles (Figure 3.2). This is in marked contrast to the brains of androgenetic chimeras which were smaller, both in absolute size and especially relative to body weight, because Ag chimeras were also of significantly larger body size than either normal or parthenogenetic embryos (Figure 3.2).

It was somewhat surprising to find that parthenogenetic cells proliferated at the expense of normal cells, and in doing so produced a larger cortex in these chimeric mice (Figure 3.2). Moreover, this enlarged brain appeared to be anatomically normal. This was even more surprising, because a large number of genes had been silenced in these cells (all of the imprinted genes that are usually paternally

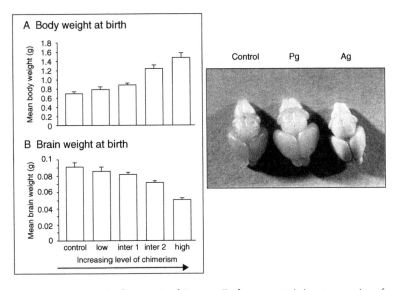

FIGURE 3.2 Androgenetic chimeras. Embryos containing two copies of androgenetic cells (Ag) have smaller brains and larger bodies compared with controls, while two copies of maternal cells (Pg) had larger brains but bodies which were of normal size, but smaller than Ag chimeras (body weight for females not shown). Increasing levels of Ag chimerism produced increasing body and decreasing brain weights.

expressed), while the autosomal genes that are maternally expressed had been duplicated. Neurological findings of human brain dysfunction seen in the Prader–Willi and Angelman's Syndromes are produced when duplicate genetic copies of restricted regions on either the maternal or paternal human chromosome 16 takes place, findings which are also congruent with the experimental mouse chimeric cell findings. Angelman's Syndrome is characterised by epilepsy, sleep disorders, ataxia and tremor, as well as speech disorders. Neurocognitive impairment is always present as well as psychiatric comorbidities. Prader–Willi Syndrome is characterised by obesity, overeating and neuroendocrine dysfunction.

The distinct patterning in the distribution of parthenogenetic and androgenetic cells and the differential effects of these cells on brain growth strongly suggested that genomic imprinting may have played an important role in forebrain evolution. This viewpoint is supported by comparative anatomical studies across the brains of different mammalian species. Indeed, mathematical allometric scaling for those different parts of the brain to which either maternally or paternally expressed genes differentially contribute reveals that a remodelling of the brain's structure has indeed occurred for these regions during mammalian evolution (Keverne *et al.*, 1996). Starting with the small insectivorous mammals and proceeding across the prosimian primates to large-brained simian primates, it is notable that the neocortex and striatum (information-processing executive brain) have increased significantly in size relative to the rest of the brain and body weight. In contrast, the 'emotional controlling' basal areas of the limbic forebrain have decreased in size over mammalian evolutionary progression. Such phylogenetic contrasts for brain growth across mammals are, at least in principle, generated and exaggerated by the presence of either parthenogenetic or androgenetic chimeric cells in the experimental studies on mice. Because it is primarily the imprinted genes that differ in Ag and Pg chimeras, then genomic imprinting is congruent with having facilitated a rapid, nonlinear expansion of the brain (especially the executive neocortex and striatum) relative to body size during the evolutionary developmental progression of these neural structures across mammals. It is important to stress that this neocortical expansion is not specific to the development of sexual differences in the brain, as this effect of chimeric cells is the same in both males and female embryos. However, it is the parental origin (mother or father) for expression of these autosomal genes that is determined by the imprinting process, and this, in turn, is responsible for these brain growth differences. Autosomal gene expression also explains why the effects of chimeric cells are the same in embryos of both sexes.

Unquestionably, genomic imprinting has played an important evolutionary role in mammalian neocortical brain development by enabling motivational behaviour (sexual, aggressive, feeding and maternal care) to become increasingly regulated, if not indeed dominated, by the enlargement of the executive neocortex.

In this way, much of motivational behaviour has become more dependent on experience both within and across generations in mammals with a large neocortex (Keverne, 2015). Together with the brain's ventral striatal reward system, this evolutionary trend for developmental growth of the executive brain may have been an important factor in developing intelligent behavioural strategies, and in the emancipation from deterministic effects of hormones on behaviour.

EVOLUTIONARY ORIGINS OF GENOMIC IMPRINTING

Approximately only 1.5 per cent of the mammalian genome comprises the DNA of protein-coding genes, while there is an abundance of retrotransposon DNA included in 'short interspersed nuclear elements' (SINEs), 'long interspersed nuclear elements' (LINES), and LTR (long terminal repeat) retrotransposons/retroviral elements. Around 50 per cent of human DNA is estimated to be that from transposons (Lander 2001). These transposons are responsible for a variety of mammalian-specific traits, and are thought to have played an important role in mammalian evolution, including genomic imprinting for development of the placenta. Some of the first imprinted genes are thought to have initially evolved from the domestication of retrotransposons (Kaneko-Ishino & Ishino, 2010), and early placental evolution depended on the syncitin transposon genes. Syncitin transposon genes produce the outer layer of the placenta which interfaces with the mother's uterus and promotes placental invasion of the maternal uterus. Although widespread across eutherian mammals, genomic imprinting and placentation also occur in the marsupials, but not in the egg-laying monotreme mammals (Renfree *et al.*, 2013) In marsupials the placental attachment to the mother is short-lived, and much of the early foetal growth occurs in the mother's pouch, supported by mammary gland milk. Here, in the Tamar Wallaby, genomic imprinting is found in the placenta, but also provides for imprinting of nutrient regulatory genes in the mother's mammary gland (Stringer *et al.*, 2014).

Two of the imprinted genes, which are specific to eutherian mammals, and have been derived from LTR retrotransposons, are *Peg10* and *Peg11* (also known as *Rtl1* and *Dlk1*) and both are essential for placental development. Deletion of the paternally expressed gene (*Peg10*) causes early embryonic death due to impaired placental growth, notably caused by growth defects of the labyrinthine and the spongiotrophoblast layers of the placenta. This same gene (*Peg10*) is also essential for nutrient and gaseous exchange between maternal and foetal blood cells, while *Peg11* gene ablation results in impairment of late foetal growth. Such later growth impairment is also due to abnormalities at the placental–maternal interface. Neither of these imprinted genes are present in the monotreme mammals. In total, nine other genes have been identified in mammals as containing the same Suchi-ishi retrotransposon, and eight of these genes have been located to the X-chromosome. Some of these X-chromosome retrotransposed genes have also been found to impair normal brain development (McCole et al., 2011).

Retrotransposons are 'selfish' genetic elements which have been hypothesised to advance evolution, but it is also the case that they may induce genetic errors. In mammals these include errors that affect brain development, such as those found in Rett syndrome, and in the context of somatic development, their induction of genetic errors are known to produce cancers. Thus, the evolutionary impact of retrotransposons has also depended on the genome's ability to regulate the inherent instability of retrotransposons, primarily via DNA methylation silencing (Ollinger et al., 2010). This is mainly brought about in the early post-fertilisation period of zygotic development by the maternal genome's expression of the *Trim28* and *Zfp57* genes (Figure 3.3) (Turelli et al., 2014). These two genes regulate epigenetic stability and protect maternal imprints from demethylation, thereby safeguarding the transcriptional dynamics of early embryos. The resistance of retrotransposons to demethylation can lead to their transgenerational inheritance, and in certain cases these have led to the parent of origin methylation imprints seen in genomic imprinting

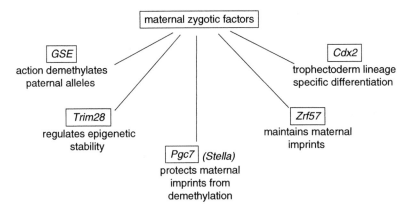

FIGURE 3.3 Zygotic methylation reprogramming. Following fertilisation and paternal DNA mismatch repair, maternal zygotic factors ensure stability of maternal imprints, demethylate maternal alleles, sustain maternal imprints, and activate trophectoderm to form the placenta.

(Suzuki *et al.*, 2007). To date, very few genes have been shown to be directly imprinted in this way, although recent evidence suggests that an abundance of genomic microRNAs, which also play an important role in genomic imprinting, have been found to be initially formed from transposable elements (Roberts, 2015). A number of studies in mammals have also found phenotypes brought about by the silencing of transposable elements, such as that which occurs in the Agouti mouse. Here, transcription resulting from a retrotransposed insertion upstream of the Agouti gene causes ectopic expression of the Agouti gene, resulting in yellow fur, obesity and diabetes (Daxinger & Whitelaw, 2012). This same genetic locus has been shown to be under rapid adaptative selection for coat colour in mice.

The elevation of cellular retrotransposons is also thought to induce meiotic prophase defects, including oocyte aneuploidy and, as a consequence, potential embryonic lethality. It is well established that death of female oocytes, but not male spermatozoa, is most likely to take place at the early meiotic prophase, causing high levels of attrition. This may ensure that only healthy oocytes survive for development of the next generation. There are but a few 'imprints' which have originated in the male germline during spermatogenesis, and these paternal imprints have a very different evolutionary history. Such male imprints have originated by retro-transposition from alien viral DNA, which effectively silences paternal DNA gene expression, thereby endowing maternal DNA with the responsibility for the functions which these genes perform. It is not, therefore, the selectivity of retroviral DNA for male or female germlines that has progressed mammalian evolution, but how this is employed by the host cell. In the matriline it would appear that retroviral insertions serve for selection of healthy oocytes, while in the male germline these insertions may be used for silencing paternal gene expression, thereby allowing selection pressures to operate for these genes primarily via the matriline. Retroviral DNA can be dangerous and is usually epigenetically silenced by a mechanism which may also extend to silencing the somatic expression of autosomal genes in their close vicinity.

In both types of imprinting, be it matrilineal or patrilineal, the mechanism for gene silencing has progressed across mammalian evolution and has seen the recruitment of more autosomal genes to these heritable epigenetic marks (Wolf, 2013). Hence the number of genes regulated according to parent of origin has expanded from nine in the marsupials, the earliest placental mammals to evolve, to approximately 150 in humans. These genes are important regulators of development and are located in some 16 imprint-controlled clusters across the various autosomal chromosomes. The clustering of genes to an ICR ensures their monoallelic expression for providing the tight control of gene dosage for in-utero development.

While less than 2 per cent of human DNA is translated into the production of proteins, the vast majority is transcribed into producing various types of non-coding RNA, including long non-coding RNA (lnc-RNA) and short micro-RNA (miRNA). The lnc-RNAs have developmental and tissue-specific expression patterns that are able to regulate neighbouring gene expression in *cis*, or more distant genes in *trans*. Some lnc-RNA constructs have been shown to bind polycomb repressive complexes, inducing epigenetic changes to histone marks and chromatin structure, ultimately suppressing transcriptional activity. Other lnc-RNAs serve as precursors for miRNAs that regulate genomic imprinting. Although the ICRs are almost exclusively maternal, the paternal gene expression that follows is nevertheless under tight control of the maternal epigenome. Thus of the 10 genes which define the paternal imprinted *Dlk–Dio3* cluster, seven code for non-coding RNAs with maternal expression. Likewise, the H19 lnc-RNA can serve as a precursor for miRNA (miR-675), which responds to hypoxic stress during development. H19 is maternally expressed in the placenta at a time when placental growth is almost complete. The physiological role for miR-675 is therefore thought to prevent placental overgrowth. Thus, placental growth is promoted by H19 regulation of Igf2, and curtailed towards the termination of pregnancy by miR-675.

What was the biological context for shaping the evolution of imprinted gene expression, and what are its advantages? Imprinted genes have undoubtedly played a major role in mammalian evolution, reaching a pinnacle of functional complexity, especially for the placenta and brain in humans. Many of these imprinted genes are co-adaptively expressed in the developing placenta and also in the developing brain (Table 3.1, Figure 3.4) (Keverne, 2015). From a functional viewpoint, mammalian in-vivo placental development, and all the advantages this entails for the foetus (warmth, food and oxygen transfer), is the primary responsibility of mother. Likewise, following birth, maternal feeding of the newborn foetus (lactation), warmth for the neonate via maternal nest-building, infant protection and maternal care are also the primary responsibility of the mother. Thus, 'in-utero' mammalian evolutionary success has depended principally on co-adaptation of the mother's brain and the foetal placenta. Hence the selection pressures which have determined the mechanisms for achieving this success are also primarily under maternal control. Because both sons and daughters benefit equally from placental viviparity, and the mother benefits equally from the birth of sons and

Table 3.1 *Maternally imprinted genes expressed in hypothalamus and placenta.*

Zac1	(P)	Hypo	Cell cycle arrest/apoptosis
Necdin*	(P)	Hypo	Growth suppressor/apoptosis
Dlk1	(P)	Hypo/pit	Cell cycle arrest/apoptosis
Sgce	(P)	Hypo	Sarcoglycan family
Peg3*	(P)	Hypo/pit	Bac transport – apoptosis
Peg1*	(P)	Hypo/pit	Growth regulation
Magel2*	(P)	Hypo	MAGE protein/apoptosis
Grb10	(P)	Hypo	Apoptosis
Nnat	(P)	Hypo	Phosphorylates CREB ↑ Ca^{2+}

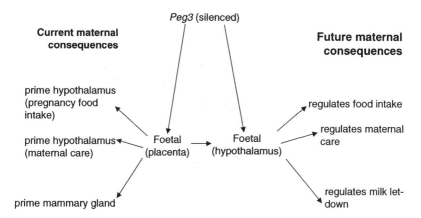

FIGURE 3.4 Genomic imprinting co-adapted functions. The consequences of silencing the *Peg3* imprinted gene in the placenta, and independently in the hypothalamus produces remarkably similar phenotypes. This illustrates the importance for genomic imprinting for brain and placental co-adaptation.

daughters, genomic imprinting has evolved to secure and fulfil this viviparous evolutionary developmental role (Renfree *et al.*, 2009).

EVOLUTIONARY THEORIES OF GENOMIC IMPRINTING

At the time when the very early chimeric studies were carried out in 1995, the understanding of genomic imprinting was still in its infancy and few imprinted genes had been identified. However, this did not prevent evolutionary biologists from developing a theory of 'parental conflict' (Haig & Graham, 1991). They hypothesised that the growth factor (Igf2 expressed from the paternal genome) and its receptor (Igf2r from the maternal genome) represented a 'parental tug-of-war'. This was translated as the father's genome operating to develop larger offspring, which was hypothesised to be in 'conflict' with the mother's genome. Mothers preferred a more balanced strategy for offspring growth and development, in line with their capacity for successful nurturing of these and future babies (Haig, 1992). However, babies that are too large are a risk to the mother's survival and hence to their own survival, while babies that are too small are only a risk to infant survival. Perhaps a better way of conceptualising these particular imprinted gene actions is to consider such early imprinted genetic events, which occur during in-utero development, as being co-adaptive. They serve as a 'rheostat' to optimise a balanced foetal growth that is in the interests of both mother and offspring. Compared with the mother, the father has little to lose in the biologically strategical sense, other than his investment in mating. The time spent mating is incomparable to the lengthy time and the energy investment made by the mother during pregnancy, lactation and parental care. Moreover, imprinting and parental conflict does not make sense when the control over imprinting for these genes is maternal, and any impairment to in-utero growth, or to placental development, is a high risk for survival of both mother and foetus. Other theories based on co-adaptation across parent offspring genomes through genomic imprinting have gained more credence in recent years (Keverne, 2015; Wolf, 2013).

Theories of genomic imprinting have tended to focus on the genes that are expressed as a consequence of the imprinting mechanism. Many of these genes are not new, but are established genes which only became imprinted with the evolution of the placenta and maternal viviparity. Moreover, most evolutionary studies have failed to take into account the epigenetic imprinting mechanisms for regulating these established genes. The ICRs for controlling genomic imprinting are primarily maternal, and ICRs are also reprogrammed in the female germline. Indeed, the epigenetic mechanism for maternal imprinting has itself become heritable. The exceptions are the *Rtl1* and *Dlk1* paternally expressed genes, the imprinting of which occurred after the marsupial to eutherian mammalian divergence, when the silencing of their maternal alleles was brought by the accumulation of mi-RNAs on the maternal chromosome.

Maternal H19 regulates the imprinting of the Igf2 locus for placental and embryo growth and is also a non-coding construct made up of multiple mi-RNAs transcribed from the maternal chromosome (Bergman et al., 2013). Multiple levels of control for expression of the paternal Igf2 locus thus involve the maternal genome, which is difficult to reconcile with this genetic locus as supporting a mother–infant 'conflict theory'. Moreover, because the action of paternally expressed Igf2 on the maternally expressed Igf2 receptor was integral to the conflict theory, this would require the maternal genome to be in internal conflict. The expression of these imprinted genes is better considered in the broader context of balanced regulation of in-utero foetal growth during pregnancy. Importantly, this balanced growth is primarily under control of the maternal genome. Towards the later stages of pregnancy, the balanced regulation of foetal placental growth is a function of the maternal H19 locus itself. The presence of a micro-RNA (Mir-675) embedded in the first exon of H19 and expressed exclusively in the placenta was subsequently identified in this context (Keniry et al., 2012). This non-coding RNA specifically targets growth of the placenta where it is rate-limiting for placental growth, thereby preventing placental overgrowth in the later stages of pregnancy. This kind of finding reflects more on a balanced co-adapted mother and foetal developmental process.

Co-adaptive gene expression in the context of viviparity and in-utero growth of the foetus is clearly essential for the mother's reproductive success, but also raises the question as to why most of the epigenetic imprints are maternal and therefore result in paternal expression of the genes regulated by maternal ICRs. The answer to this question lies in the important role which the matriline plays in imprinted gene stability.

I have described in some detail how the mammalian matriline has taken the lead in advancing mammalian evolution at the genetic and epigenetic levels, and how genomic imprinting has been particularly effective in achieving this success. In considering evolutionary success at the biological level, it was Trivers and Willard (1973) who drew attention to the advantages derived from the different reproductive strategies adapted by males and females for achieving reproductive success. Females enhance their reproductive success by producing viable offspring through the provisioning and transfer of maternal energy resources during pregnancy and post-partum nursing, together with maternal caring. In contrast, males optimise their reproductive success by finding reproductively fertile females and mating with as many as possible. Both strategies have been shown to be under the regulatory control of imprinted genes (Swaney et al., 2007).The males of most mammalian species optimise their reproductive success via their chemosensory system responding to oestrous pheromone signals, while the female produces viable offspring by responding to hormonal signals from the foetal placenta.

In considering the important role for imprinted genes in mammalian reproductive success, the question arises as to why the female germline carries out imprint silencing of her allele, thereby enabling only the paternal to be expressed (Figure 3.5). Paternal expression of the maternally imprinted gene enables faster propagation of the allele throughout the population, while the female's silent copy provides a stable template for DNA repair during homologous recombination. Moreover, by ensuring paternal expression, the matriline employs spermatogenesis as a 'driver' for diversification and variance of paternal downsteam genes. Matrilineal control over genomic imprinting thereby deploys the patriline for rapidly establishing the gene in the population, and for the reproductive success that is provided by male downstream genes, while the matriline provides stability for those important and stable regulatory genes that are imprinted.

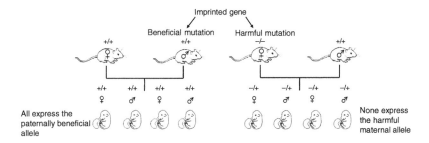

FIGURE 3.5 Inheritance and expression from maternal imprinting giving rise to paternal gene expression. Offspring carrying a beneficial mutation to a paternally expressed gene from the father pass this on to all of their offspring. Should a gene be disadvantageous when inherited from the matriline, it can undertake mismatch repair by the matriline.

Additional biological theories for the advancement of mammalian evolution include 'niche construction' where attention has been directed to environmental modifications that are constructed by the animal and that shape its reproductive success (Laland *et al.*, 2016). Examples of this include the Beaver's water dam and the rabbit's underground burrow. In these examples, the mammals' survival and their evolutionary trajectory is advanced by the animal itself. There are similarities between this theory and Dawkin's concept of the 'extended phenotype', which is based on animal adaptations whose genes underlie construction of the trait. For all mammalian species it might be of evolutionary help to think of parental construction of the placenta and foetus as part of a different generation that is adapted to developing within the mother's uterus. This may be viewed as a specialised niche which has shaped mammalian evolution across generations, and in this context we know a great deal about the genes, many of which are imprinted (Keverne, 2015). Indeed, the foetus influences its own evolution by being both the object of natural selection and the creator of the conditions for that selection (Lewontin & Levins, 2000). Imprinted genes have indeed played a prominent role in the transgenerational co-adaptation between mother and foetus (see Chapter 5).

IMPRINTED GENE STABILITY

Although genomic imprinting results in alleles that are either maternally or paternally expressed, it is the matriline which takes the genomic lead in determining which of these parental alleles are expressed. We might therefore ask, if the mother is really in control, then why does the maternal genome sometimes arrange for the father's potentially more 'mutable' alleles, arising from male spermatogenesis, to be expressed? In evolutionary terms the mother can only afford to permit such an event providing any mutated paternal alleles are securely repaired and stabilised. Should mutations occur, advantageous or otherwise, the mother needs to be certain they are repaired to match her own silent but stable imprinted copy of this gene. The evolutionary terminology for this kind of gene stabilisation is 'purifying selection', rather than 'natural selection'. Purifying selection through maternal mismatch DNA repair is an integral part of paternal gene regulation, recovering and stabilising the paternal alleles that arise from the mutagenic environment created by the multiple cell divisions that are a part of spermatogenesis (Keverne, 2014).

The imprinted genes do indeed have remarkable stability, which is counter-intuitive to our views on the key role which they have played in mammalian evolution (Hutter *et al.*, 2010). None of the imprinted genes have followed the usual evolutionary trajectory of gene duplication or changes to the protein sequence for which they code. These imprinted genes actually display high levels of conservation in their coding sequence across different placental species. In non-placental mammals, as well as in reptiles and frogs, where many of these same genes persist but are not imprinted, they have in their past phylogenetic history been subject to dynamic evolutionary changes in structure. Thus, in placental mammals these same genes, once they became imprinted, have subsequently undergone the process of 'purifying selection', making them the most stable genes in the human genome. This would suggest that once natural selection has found a successful template for stabilising what is a potentially 'risky in-utero form of development', then evolution has capitalised on imprinted genetic stability. Evolutionary variance is primarily secured via the downstream secondary and tertiary networks of the imprinted genes (Varrault et al., 2006). It is these large genetic networks (Figure 3.6), engaged by the imprinted genes, that produce much of the variance across species. Moreover, it has been in the matriline's interest to imprint genes for paternal expression, thereby enabling a more rapid spread and subsequent establishment of these imprinted genes in the population. It has also been demonstrated from gene 'knockout' studies that genomic imprinting improves the male's ability to find oestrous females and secure fertile mating (Swaney et al., 2007). Such genetically advantaged males can thus spread these advantages derived from genomic imprinting more rapidly throughout the population by mating more females and siring more offspring. In contrast, the advantage accruing to females from

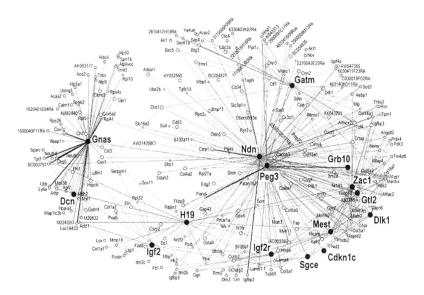

FIGURE 3.6 *Peg3* belongs to a network of correlated imprinted genes linked to other imprinted genes. The interactions of gene networks that employ imprinted genes at their hub are clearly complex and interlocking. (With permission from Elsevier Science, Varrault et al., 2006). Imprinted genes are shown in bold type.

imprinted genes ensures that each female cares for and successfully rears these advantaged offspring.

Although many of the same genes existed prior to imprinting and exhibited the capacity for mutational change in reptiles and early non-placental mammals, on becoming imprinted in placental mammals they actually became buffered against such mutational change. As already mentioned, their stability can be explained by a DNA repair mechanism that takes place during the germline reprogramming of the imprints for the next generation (Wolf & Hager, 2006). This repair mechanism is designed to correct the genetic errors that are known to occur more frequently in the production of sperm. Errors in male DNA are particularly notable during spermatogenesis and occur as a consequence of the multiple cell divisions during the production of thousands of spermatozoa. Each sperm stem cell undergoes six cell divisions producing 64 primary spermatocytes, which subsequently result in the production of around 300–600 sperm per gram of testes per second! The more frequent errors that inevitably occur in spermatogenesis are primarily due to the expansion of repeat sequences that disrupt coding, and which may also induce structural chromosome changes. These errors have all been documented as occurring more frequently through paternal transmission of non-imprinted genes in humans (Marchetti *et al.*, 2007). However, this is not the case for imprinted genes that are expressed from the paternal chromosome, because sperm DNA undergoes repair immediately after fertilisation in the zygote. This occurs after the sperm nucleus fuses with that of the oocyte, and hence comes under the control of the mother's germline. The male and female genomes at this early zygotic stage are very dissimilar, with male chromatin undergoing widespread demethylation under matrilineal control, and it is this process which is linked to the DNA mismatch repair mechanism. Hence, DNA repair during this early developmental period can be regarded as an important maternal trait for repairing genetic errors that have arisen on the paternal chromosomes (Wossido *et al.*, 2010). Genes that play an important role for the success of in-utero development require stability to avoid any potentially harmful mutations for development of the foetus. These, in turn, could have a knock-on effect for survival of others in the litter, and hence for the mother herself.

There are several other features of the matriline that sustain and control the early period of post-fertilisation development (Hatanake et al., 2013). Sperm DNA is tightly packaged by proteins (protamines), a process which ensures minimal nuclear volume. This small, compact phenotype makes the sperm capable of swimming long distances relative to their size. Although their tightly protein-packaged DNA facilitates sperm motility, in such a compact state paternal DNA is consequently not available for transcriptional expression immediately after fertilisation.

A further important cellular organelle is the mitochondrion, which provides the cell's energy. However, the propulsion energy required to enable sperm to reach the female cervix and swim the long distance to the ampulla of the fallopian tube, where fertilisation occurs, severely depletes mitochondrial energy. Exhausted of their energy-producing capabilities, male mitochondria play no further part in development and are rejected by the egg at fertilisation. This essential energy-producing role in the fertilised egg is therefore taken over exclusively by the maternal mitochondria. Albeit complex, we thus have a strong matrilineal control over the earliest stages of post-fertilisation DNA repair, which allele of the gene is expressed, and the fertilised egg's capacity for energy production. In this way the subsequent engagement of maternally imprinted genes is able to optimise error-free mammalian development.

IMPRINTED GENE NETWORKS

There has been a progressive recruitment of more genes to the imprint control regions throughout mammalian evolution. Not only do these genes share the same imprint, but they tend to network with other imprinted genes and with additional downstream genes (Varrault et al., 2006) (Figure 3.6). This has been revealed from single imprinted gene 'knockout' studies in mice, as well as from the complex syndromes that arise from 'loss of imprinting' in human disorders such as Prader–Willi syndrome (Cassidy et al., 2012), and Angelman's syndrome (Thibert et al., 2013). There is a popular belief among lay people that there is a 'gene for this function' or a 'gene for that behaviour', but in reality the 'this and that' are coded for by a multiplicity of genes which function both within and across genetic networks. Moreover, a given gene in this network can, in other tissue contexts, engage with genes across other networks. The recent 'Encode' investigation of the human genome, carried out by many genome laboratories around the world, has found transcription factors that initiate expression of maternal-specific networks (Gerstein, 2012). Some 4800 specific transcription factors have been identified as mapping to and engaging maternally regulated target genes. As the degree of combinatorial regulation increases, so does the relationship between the activator gene and the activated gene become strengthened. Thus, many of the target genes that are regulated by maternal transcription factors also show high levels of maternal gene expression, but this is not the case for either paternal alleles or for paternal transcription. It could be argued that the mother already has control over the safety of her own alleles, while many of the paternal alleles have already been stabilised through maternal imprinting and the purifying selection that has accompanied this.

In addition to the networks across and between imprinted genes, each imprinted gene may itself regulate a network of downstream genes engaged in specific functions. As an example, the imprinted *Peg3* gene transcriptome has a network of 22 genes concerned with neural development, and 21 genes that are regulating other transcription factors. *Peg3* also regulates a plethora of genes concerned with placental development (Figure 3.7). It is these downstream genes that have undertaken multiple duplications that differ across different mammalian species, with the lowest copy number in the early metatherian mammals (marsupials) and the highest copy number in the more advanced eutherian mammals (monkeys and apes). Moreover, a large number of these imprinted genes are

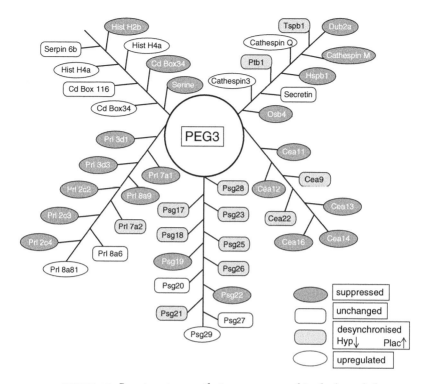

FIGURE 3.7 *Peg3* target genes that are coexpressed in the hypothalamus and placenta on days E11–E12. Effects of silencing *Peg3* on this gene network of coexpressed genes in the placenta and the female brain.

coexpressed in the developing placenta and in the developing brain, providing the potential for intergenerational coadaptive functioning of these structures (see Chapter 5). The stability of imprinted genes and the robustness of their functional networks thus allows for their interacting hubs to provide for compensatory actions via their common downstream genes in the development of such fundamentally important mammalian organs (placenta and brain). Thus, any minor genetic errors that occur may be absorbed and compensated for through the imprinted gene's network, enabling developmental robustness to prevail. In this way mammals avoid the potential risk to the mother from developmental in-utero errors, and hence avoid the potential for failure of this and future pregnancies. Moreover, the substantial contribution of the same imprinted genes to development of the hypothalamic brain and placenta has also been of considerable significance to the co-adaptive evolution of these structures (see Chapter 6).

REPROGRAMMING OF THE IMPRINTED GENOME

Human development begins with the bringing together of male and female gametes to form the fertilised zygote which subsequently develops into some 200 histologically distinct cell types. These different cell types each contain the same set of genes. As each of these 200 cell types are genetically identical to the original fertilised zygote, it should theoretically be possible to wind back the programme of gene expression in any cell and create a new, albeit genetically identical, individual. Indeed, this is precisely what Sir John Gurdon achieved way back in 1962 when he removed the nuclear DNA from an adult differentiated epithelial cell in the frog and inserted this nucleus into an enucleated egg of the frog (Gurdon, 1962). This procedure successfully created a new generation of tadpoles. For this work Professor Gurdon shared the Nobel Prize in 2012 with Professor Yamanaka from Japan, who identified four mammalian genes, the expression of which provides the key to mammalian genomic reprogramming (Yamanaka, 2013). These studies demonstrated the availability of germ cell mechanisms that are capable of allowing differentiated cells to perpetuate the molecular genetic memory for those developmental decisions that originally created it.

When the earliest stages of mammalian development were first investigated, it was found that the epigenetic structure of DNA which ensures gene silencing contrasts markedly with DNA structure when this is available for transcription. To understand how this awakening of the genes is brought about, some knowledge of the biochemical changes to DNA structure is required (Deaton & Bird, 2011). These biochemical changes either close DNA down and produce gene silencing by methylation of the regions that promote gene expression (the condition found in the sperm nucleus), or make DNA available for expression by demethylation (the process that occurs in female oocyte development and post-fertilisation). A number of specific enzymes are available for each of these processes. The zygotic DNA from sperm in the fertilised egg is comprehensively methylated, while that of the female oocyte is only moderately methylated. Thus some 1600 regions of DNA are differentially methylated between sperm and oocytes, and this differential methylation localises to different sites on DNA. Because DNA methylation suppresses gene transcription, its erasure by demethylation is an essential component of genomic DNA reprogramming to a state of totipotency. As early embryo development proceeds to implantation in mother's uterus, some cells of this blastocyst have already acquired tissue-specific identities (trophectoderm that becomes the placenta, and inner cell mass that forms the embryo). These developments are accompanied by increases in specific gene re-methylation. Thus, after total erasure of the parental epigenetic memory, a new epigenetic memory for the foetal genome is progressively restored.

Important to this process of reprogramming is the resetting of 'imprints' that enable the next generation to once again express specific autosomal genes according to the parent of origin. Without the removal and replacement of these imprinted epigenetic marks, the offspring of future generations would fail to appropriately express the key developmental genes which these imprints control. The maternal ICR is only effective when it takes its origin in the matriline. Thus, germline reprogramming of imprints (erased from the sperm of the next generation) is essential to ensure matrilineal continuity. Only daughters have the capacity to carry forward this 'imprint' to the following generation. Interestingly, the genomic reprogramming of germ cells also differs according to whether they are male or female germ cells. In females, this takes place in the developing foetal germline of the next generation (around weeks 5–7 of pregnancy in humans) when the foetus is in the mother's womb. In males there is substantial reprogramming of male DNA in sperm, but this takes place in the adult male only days before fertilisation, and a second reprogramming of paternal DNA takes place post-fertilisation with assistance from the matriline. In contrast, the maternal imprints are reprogrammed in the female oocytes which are developing in the foetal ovaries while the foetus is still *in utero*, a generation ahead of male germ cell reprogramming. Maternal imprints are thus established a generation ahead of male reprogramming and hence require safeguarding throughout the second phase of post-fertilisation reprogramming of the paternal genome. Again, it is maternal factors present in the fertilised egg which protect the maternal imprints while the maternal post-fertilisation reprogramming of the father's epigenome takes place (Figure 3.3).

It is important to note that in all of the experimental reprogramming studies, the adult cell nucleus is transferred to the enucleated maternal oocyte and not to the paternal sperm. The sperm's DNA is virtually unmodifiable and incapable of expression due to the tight packaging of its DNA with proteins (protamines). Hence, sperm DNA further requires epigenetic assistance from the maternal oocyte after fertilisation in order to initially ensure removal of the sperm protamines and their replacement with histones. Histones provide a more open chromatin structure, enabling the paternal genome to renew gene transcription. Thus, early development of the fertilised egg differs in a number of respects for the maternal and paternal genomes. It is primarily the maternal alleles of the fertilised egg that are exclusively expressed at the very earliest post-fertilisation stages of embryo development. Indeed, these maternal zygotic factors are essential to enabling the expression of paternal alleles. Zygotic *Gse* demethylates paternal alleles enabling their expression (Hatanake *et al.*, 2013), while zygotic *Pgc7* and *Stella* protect and sustain maternal imprints from methylation removal during this post-fertilisation period of male germline reprogramming (Figure 3.4) (Bian & Yu, 2013). Equally important is the release of the paternal chromatin from its tight packaging in readiness for initiation of the mismatch repair of the sperm DNA, thereby eliminating minor paternal genetic errors (Hajkova *et al.*, 2010). Importantly, it is the early expression of maternal alleles that initiate the onset of placental development and it is the maternal mitochondria that provide the energy for the post-fertilised egg development. Thus, although both male and female genomes are essential for normal development, at this vital development time it is the matrilinc which takes the leading role.

Although reprogramming of both male and female genomes is essential for the successful development of each new generation, in Placental mammals the matrilineal genome takes the lead in the reprogramming of imprints and in enabling the initial transcription of the male genome. The earliest development decisions to be made by the fertilised oocyte are regulated by the mother's genome, and the timing for the onset of paternal gene transcription depends initially on maternal transcription. One of the first genes to be expressed (*Cdx2*) codes for the determination of trophectoderm, the tissue which progresses to become the placenta, and this too is a maternally expressed gene. Some 15 chromosomal loci that exhibit parent-of-origin methylation are restricted to the placenta (Monk, 2015). The process of reprogramming is complex, but it is integral to understanding genomic imprinting and mammalian development, which in turn has been integral to the reproductive success of mammals.

Progress in our knowledge for this area of developmental research is moving very fast, and it was only recently that new findings for reprogramming of the human genome were discovered (Tang *et al.*, 2016). A number of human genes actually escape reprogramming in the female germline and remain silent until activation of the genetic programme for brain development is called into action. Human brain development, especially of the executive neocortex, continues throughout adolescence until the age of 22, but it is not yet known exactly when these genes, which escape early demethylation, undertake their phase of reprogramming and become active in the brain. It would make good sense to conserve such genes exclusively for expression in the brain as it provides these 'escapee' genes with a unique commitment to cortical executive brain development. They can therefore be subject to, and specialised for, selection pressures specific to the brain's later development. In this way, those 'escapee' genes which are selective for brain development are not required to contend with potentially conflicting commitments and selection pressures during the earlier stages of development, should they have first been expressed in other tissues.

There are many clinical disorders that are associated with the large number of imprinted genes that are expressed in the brain. Like many genetic disorders, some of these brain dysfunctions are associated with mutations to the imprinted gene that is expressed. Recent advances in the technology that enables the investigation of genomic imprinting has led to the suggestion that imprinting disorders may be amenable to clinical interventions that rescue the disorder by reactivating the unaffected repressed imprinted allele. In the context of imprinted gene expression, many of these disorders are not associated with the genes themselves but with the ICR (methylation failures underpinning the ICR). Failures in methylation silencing can produce increased gene dosage (biallelic expression) or decreased gene dosage if the ICR failure normally instructs a gene that is itself a repressor (lnc-RNA, polycomb repressor complexes, or histone methyl transferases). Many of the imprinted gene clusters that have been investigated in human clinical disorders are extremely complex with a multitude of genes and gene regulators coming under the influence of the ICR. Putting theory into practice at the level of reversing genetic dysfunction is not a simple matter and one that leaves me less than optimistic.

In this chapter I have focussed on genomic imprinting from the female perspective because the imprint control regions (ICRs) that enable genomic imprinting are primarily matrilineal. The few documented ICRs that are paternal have their evolutionary origins in retrotransposed viral DNA and non-coding miRNAs. The female imprints that are found on most of the chromosomes are inherited from the mother, and the evolutionary trend across mammals has been to recruit more genes to these ICRs and thereby ensure their monoallelic expression according to parent of origin. From an evolutionary perspective a considerable amount of confusion has been generated by focussing on the gene that is expressed in genomic imprinting rather than the lead taken by the maternal ICR that underpins the mechanism allowing only this single allele to be expressed. Thus a paternally expressed allele of the imprinted gene achieves this status by virtue of the mother silencing her allele. Of course we may ask the question as to why the female should silence her allele and allow the paternal allele to be expressed. Herein resides the mechanism for explaining the evolutionary success of genomic imprinting.

To address this question we first need to consider the specific developments that have taken place and have been integral to the evolutionary success of mammals. Most notable has been the evolution of in-vivo placentation together with evolution of the large mammalian brain (see Chapter 5). Certainly, most of the imprinted genes are expressed in the brain or in the placenta, and a substantial proportion of these genes are expressed in both structures. There is no question that viviparity involves a considerable investment of female time and energy to mothering. Pregnancy failure thus represents a substantial cost to maternal investment, and potentially to the mother's life. By providing monoallelic gene expression through genomic imprinting, the matriline provides tighter control over gene dosage, an integral part of the precision required for complex in-utero development. Imprinting thus ensures that it is the same genetic allele that is expressed in all of those cellular building blocks required to produce a placenta or a brain. However, this does not explain why the mother selects the paternal allele for expression when the possibility for more errors accumulating will take place during the multiple cell divisions that occur in the male germline during spermatogenesis. Such paternal errors are, however, removed and corrected by the matriline during post-fertilisation mismatch repair to ensure they genetically match to the mother's own maternal allele. In this way, monoallelic expression of imprinted genes provides tight control over both gene stability and gene dosage, both of which have been crucial to the success of mammalian viviparity.

In summary, maternal genomic imprinting of autosomal genes thus ensures genetic stability by purifying selection. Many of the imprinted genes are paternally expressed, thereby providing for faster dissemination of this stabilised paternal allele in the population as a result of the male's capacity for mating with multiple females. In support of this, some of these same paternally expressed genes function to increase the male's reproductive success by enhancing his olfactory capacity for the detection of fertile females, and pheromonally provoking arousal of his mating response capabilities. Moreover, a subset of imprinted genes are coexpressed in the maternal brain and placenta, adapting the maternal brain to respond to the hormonal messengers from the foetal placenta (see Chapter 6). A number of these coexpressed genes contain clusters of small noncoding RNAs which are regulated by the *Drosha–Dgcr8* complex, which is also expressed in the developing brain and placenta. The C19 cluster of miRNAs is associated with pregnancy complications. Different miR members are important for primary placental trophoblast development (miR141 and miR21) and for extravillous trophoblast invasion of the maternal decidua and myometrium (miR519) (Buckberry *et al.*, 2014). Others are expressed in exosomes released from the placental villous and endow non-trophoblastic cells with resistance to retroviruses (Mouillet *et al.*, 2014). In this way, the large C19 microRNA cluster is targeted by *Dgcr8* to regulate its component microRNAs for specific placental cell development (Bellemer *et al.*, 2012).

In summary, matrilineal genomic imprinting has substantially contributed to ensuring the reproductive success of male mating, to determining the success of in-vivo placentation, and to ensuring the mother is energetically and behaviourally prepared for the birth of the offspring that follow (see Chapter 6).

REFERENCES

Allen, N. D., Logan, K., Drage, D. J., *et al.* (1995). Distribution of parthenogenetic cells in the mouse brain and their influence on brain development and behaviour. *Proc. Natl Acad. Sci. USA* 92: 10782–86.

Bellemer, C., Bortolin-Cavaille, M. L., Schmidt, U., *et al.* (2012). Microprocessor dynamics and interactions at endogenous imprinted C19MC microRNA genes. *J. Cell Sci.* 125: 2709–20.

Bergman, D., Halje, M., Nordin, M., *et al.* (2013). Insulin-like growth factor 2 in development and disease: a mini review. *Gerontology* 59: 240–49.

Bian, C. & Yu, Z. (2013). *PGC7* suppresses *TET3* for protecting DNA methylation. *Nucl. Acids Res.*, 2: 2893–905.

Bourc'his, D. & Bestor, T. H. (2006). Origins of extreme sexual dimorphism in genomic imprinting. *Cytogenet. Genome Res.* 113: 36–40.

Buckberry, S., Bianco-Miotto, T. & Roberts, C. T. (2014). Imprinted and X-linked non-coding RNAs as potential regulators of human placental function. *Epigenetics* 9: 81–89.

Cassidy, S. B., Schwartz, S., Miller, J. L., *et al.* (2012). Prader–Willi syndrome. *Genet. Med.* 14: 10–26.

Daxinger, L. & Whitelaw, E. (2012). Understanding transgenerational epigenetic inheritance via the gametes in mammals. *Nat. Rev. Genet.* 13: 153–62.

Deaton, A. M. & Bird, A. (2011). CpG islands and the regulation of transcription. *Genes Dev.* 25: 1010–22.

Gerstein, M. (2012). Genomics: ENCODE leads the way on big data. *Nature* 489: 208.

Gray, A. P. (1972). *Mammalian Hybrids*. Second (revised) edn. Slough: Commonwealth Agricultural Bureaux.

Gurdon, J. B. (1962). Adult frogs derived from the nuclei of single somatic cells. *Dev. Biol.* 4: 256–73.

Haig, D. (1992). Genomic imprinting and the theory of parent–offspring conflict. *Semin. Devel. Biol.* 3: 153–60.

& Graham, C. (1991). Genomic imprinting and the strange case of the insulin-like growth factor II receptor. *Cell* 64: 1045–46.

Hajkova, P., Jeffries, S. J., Lee, C., *et al.* (2010). Genome-wide reprogramming in the mouse germ line entails the base excision repair pathway. *Science* 329: 78–82.

Hatanake, Y., Shimizu, N., Nishikawa, S., *et al.* (2013). GSE is a maternal factor involved in active DNA demethylation in zygotes. *PLos ONE* 8: e60205.

Hutter, B., Bieg, M., Helms, V., *et al.* (2010). Imprinted genes show unique patterns of sequence conservation. *BMC Genomics* 11: 649.

Kaneko-Ishino, T. & Ishino, F. (2010). Retrotransposon silencing by DNA methylation contributed to the evolution of placentation and genomic imprinting in mammals. *Dev. Growth Differ.* 52: 533–43.

Keniry, A., Oxley, D., Monnier, P., *et al.* (2012). The H19 linc RNA is a developmental reservoir of miR-675 that suppresses growth and Igf1r. *Nat. Cell Biol.* 14: 659–65.

Keverne, E. B. (2014). Mammalian viviparity: a complex niche in the evolution of genomic imprinting. *Heredity* 113: 138–44.

——— (2015). Genomic imprinting, action, and interaction of maternal and fetal genomes. *Proc. Natl Acad. Sci. USA* 112: 6834–40.

Keverne, E.B., Fundele, R., Narashimha, M., *et al.* (1996). Genomic imprinting and the differential roles of parental genomes in brain development. *Dev. Brain Res.* 92: 91–100.

Laland, K., Mathews, B. & Feldman, M. W. (2016). An introduction to niche construction theory. *Evol. Ecol.* 30: 191–202.

Lewontin, R. & Levins, R. (2000). Let the numbers speak. *Int. J. Health Serv.* 20: 873–77.

Marchetti, F., Essers, J., Kanaar, R., *et al.* (2007). Disruption of maternal DNA repair increases sperm-derived chromosomal aberrations. *Proc. Natl Acad. Sci. USA* 104: 17725–29.

McCole, R. B., Loughran, N. B., Chahal, M., *et al.* (2011). A case-by-case evolutionary analysis of four imprinted retrogenes. *Evolution* 65: 1413–27.

McGrath, J. & Solter, D. (1984). Completion of mouse embryogenesis requires both the maternal and paternal genomes. *Cell* 37: 179–83.

Monk, D. (2015). Genomic imprinting in the human placenta. *Am. J. Obstet. Gynecol.* 213: S152–S62.

Mouillet, J. F., Ouyang, Y., Bayer, A., *et al.* (2014). The role of trophoblastic microRNAs in placental viral infection. *Int. J. Dev. Biol.* 58: 281–89.

Ollinger, R., Reichmann, J. & Adams, I.R. (2010). Meiosis and retrotransposon silencing during germ cell development in mice. *Differentiation* 79: 147–58.

Renfree, M. B., Hore, T. A., Shaw, G., *et al.* (2009). Evolution of genomic imprinting: insights from marsupials and monotremes. *Annu. Rev. Genomics Hum. Genet.* 10: 11.1–11.22.

Renfree, M. B., Suzuki, S. & Kaneko-Ishino, T. (2013). The origin and evolution of genomic imprinting and viviparity in mammals. *Phil. Trans. R. Soc. B* 368: 20120151.

Roberts, T. C. (2015). The microRNA machinery. *Adv. Exp. Med. Biol.* 887: 15–30.

Stringer, J. M., Pask, A. J., Shaw, G., *et al.* (2014). Post-natal imprinting: evidence from marsupials. *Heredity (Endinburgh)* 113: 145–55.

Surani, M. A., Barton, S. C. & Norris, M. L. (1984). Roles of paternal and maternal genomes in mouse development. *Nature* 311: 374–76.

Suzuki, S., Ono, R., Narita, T., *et al.* (2007). Retrotransposon silencing by DNA methylation can drive mammalian genomic imprinting. *PLoS Genet.* 3: e55.

Swaney, W. T., Curley, J. P., Champagne, F. A., *et al.* (2007). Genomic imprinting mediates sexual experience-dependent olfactory learning in male mice. *Proc. Natl Acad. Sci. USA* 104: 6084–89.

Tang, W. W., Kobayashi, T., Irie, N., *et al.* (2016). Specification and epigenetic programming of the human germ line. *Nat. Rev. Genet.* 17: 585–600.

Thibert, R. L., Larson, A. M., Hseih, D. T., *et al.* (2013). Neurological manifestations of Angelman syndrome. *Pediatr. Neurol.* 48: 271–79.

Trivers, R. L. & Willard, D. E. (1973). Natural selection of parental ability to vary the sex ratio of offspring. *Science* 179: 90–92.

Turelli, P., Castro-Diaz, N., Marxetta, F., *et al.* (2014). Interplay of TRIM28 and DNA methylation in controlling human endogenous retroelements. *Genome Res.* 24: 1260–70.

Varrault, A., Gueydan, C., Delalbre, A., *et al.* (2006). Zac1 regulates an imprinted gene network critically involved in the control of embryonic growth. *Dev. Cell* 11: 711–722.

Wolf, J. B. (2013). Evolution of genomic imprinting as a coordinator of coadapted gene expression. *Proc. Natl Acad. Sci. USA* 110: 5085–90.

& Hager, R. (2006). A maternal–offspring coadaptation theory for the evolution of genomic imprinting. *PLoS Biol.* 4: e380.

Wossido, M., Arand, J., Sebastiano, V., *et al.* (2010). Dynamic link of DNA demethylation, DNA strand breaks and repair in mouse zygotes. *EMBO J.* 29: 1877–88.

Yamanaka, S. (2013). The winding road to pluripotency (Nobel Lecture). *Angew. Chem. Ind. Ed. Engl.* 52: 13900–09.

4 Puberty: Developmental Reorganisation of Sex Differences in Body and Mind

Puberty is the start of a turbulent period for the development of adolescence and has long been viewed as lifetime's phase of 'stress and storm'. Even dating back to the writings of William Shakespeare (1623), 'I would there was no age between sixteen to three and twenty, for there is nothing in the between but getting wenches with child, wronging the anciently, stealing and fighting' (quote from *A Winter's Tale*). You may note that Shakespeare was referring to men, but puberty is equally, if not even more, of a period for turbulent changes to the body and mind of women. What, then, is the biology that underpins puberty and, in turn, these turbulent changes? Undoubtedly the gonadal hormones, originating from the ovaries in women and the testes in men, play an important and well-recognised part in the reproductive biology of mammalian puberty, although in humans it is especially the brain which creates more turbulence, undoubtedly aided by the hormones. In what way and how, therefore, do the gonads change at puberty and why is this process delayed for so many years in humans? Moreover, the activity of the gonads at adolescence impacts differently on the behaviour of males and females; so what is happening in the brain to underpin these differences? These are all questions that need to be addressed when considering the biology of puberty.

There have been suggestions that the stress of adolescence is a product of Western societies' own construction. The American social anthropologist, Margaret Mead, found that the youth of the Samoan Islands experienced none of the rebellious or emotional behavioural problems shown by North American teenagers. She concluded these behavioural problems in American youth were the consequence of a culture that differed from that of the teenagers in Samoa (Mead, 1928).

Western cultures are undoubtedly under continuous change, even from one generation to the next, and lack the long-term sociocultural stability of the kind seen among small isolated island communities such as those of Samoa. There are, however, other differences associated with puberty in modern Western societies, the most notable being those concerned with the progressively earlier timing for the onset of puberty. In this context we need to consider the important effects of nutrition and the high-calorie processed foods available for consumption by modern Western societies. Diet is certainly a biological issue for attention in the context of early puberty onset in modern Western societies, while the stress from social pressures to achieve a successful career and to sustain teenage relationships is also of important psychological consideration throughout the adolescent period.

Adolescence is undeniably a potential crisis period for development of the brain. This is a period of considerable emotional turmoil when changes in the physical phenotype need to be matched with the reorganisational changes to areas of the executive brain, the prefrontal cortex in particular. This cortical region of the human brain is intimately concerned with control over both the regulation of emotions together with forward planning for life's future opportunities (Blakemore *et al.*, 2010). The fact that girls are likely to experience psychological problems earlier than boys may be related to the consistent findings that girls enter puberty earlier than boys. Indeed, a major change which has occurred in only the past few decades has seen an even earlier onset of puberty in both sexes which has been brought about as a result of better, or rather higher, calorie nutrition. Whether this earlier onset of puberty is responsible for the progressively earlier symptoms of psychological disturbances is also a distinct possibility, and one that needs more rigorous consideration. It is certainly the case that increased body fat seen in girls, even as young as 5 years of age, is predictive of the premature puberty which may sometimes occur at the very early age of 9 years in these girls (Davison *et al.*, 2003). Considered within an evolutionary framework,

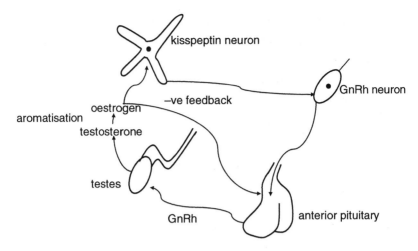

FIGURE 4.1 Kisspeptin action on gonadotrophin neurons in the pathway to gonadal hormone release. The influence of kisspeptin on gonadotrophin-releasing hormone and negative feedback via the testes steroid hormones testosterone and oestrogen. Oestrogen produces the opposite effect of kisspeptin in the male.

we can surmise that when the brain is 'made aware' that fat reserves of the body are sufficient to sustain the energy demands of pregnancy, then it is brain activity that initiates the onset of puberty. The mechanisms for how this is brought about starts with the fat cells of the body. These produce the hormone leptin, and it is this hormone which signals to the hypothalamic region of the emotional brain to activate the body's reproductive physiology (Figure 4.1) (Apter, 2003). Throughout most of human ancestry, food has been in a relative short supply for the majority of young people in the population. Hence the failure to build up sufficient bodily reserves of fat has delayed the timing of onset for reproductive activity. However, it is now the case that many Western societies are experiencing unprecedented increases in early childhood fat accumulation and, in turn, very early puberty onset. Although the human body and brain have evolved in synchrony for many thousands if not millions of years, this synchrony now appears to be failing, especially in Western societies. We might therefore ask as to how this is brought about, and what might be the psychological consequences that result from a premature onset of puberty? Are these psychological consequences due to the maturation of the body's phenotype taking place out of phase with the executive brain's development? Is it the body's precocious fat reserves that are responsible for creating many of the psychological problems that occur among present-day adolescents? Alternatively, is it the body's fatty phenotype which is itself prematurely precipitating the emotional brain's maturation out of phase with the more rational executive brain's development? It seems likely that each of these events could be an issue for the well-being of modern-day teenagers, and may indeed underpin the human brain's vulnerability to an increasing predisposition for psychological disorders during the adolescent period (Frombonne, 1998a).

The gonads of males and females (testes and ovaries), when they undertake maturation at puberty, start the production of the sex-dependent hormones. These in turn bring about the all familiar, but sexually different, anatomical changes in reshaping the body's maturation throughout adolescence. However, the signal for these hormonal changes is under the control of the brain. More specifically, it is those parts of the phylogenetically older emotional brain, a region of the brain that is common to all mammals and named the hypothalamus, which primarily responds to the fatty hormone (leptin) signals from the body. The hypothalamus is itself located in the basal part of the forebrain, situated below the thalamus (hence its name), which in turn is encased by the cortex. The neural substructures of the hypothalamus have an intimate role in the regulation of primary motivated behaviour (sex, aggression, feeding and parental care), and these neural based behaviours are brought about by the action of, and feedback from, the body's various hormones. The regulatory synchronisation of this part of the emotional brain with the body's sex hormones, which start to be secreted by the gonads at puberty, is fundamental to reproductive success. The hypothalamic regions of the emotional brain achieve this successful synchronisation of brain and body via a small 'pea-like' structure, the neuroendocrine pituitary gland, which sits below the hypothalamus. This master gland of the neuroendocrine system interfaces and controls many of the body's endocrine organs, the most notable of which, in the context of puberty, are the ovaries and testes. These downstream endocrine glands produce hormones (oestrogen and progesterone in females, and testosterone in males) which in turn synchronise brain/body activity (Keverne, 1985). There are two anatomically distinct regions to the pituitary gland. The posterior part arises during development as a ventral outgrowth of the brain's hypothalamus, and contains the nerve terminals for the release of the hormones oxytocin and vasopressin, neuropeptides that are important for successful pregnancy. The anterior part of the pituitary gland also secretes a variety of peptide hormones directly into the bloodstream, and these regulate the production of steroid hormones produced by the ovaries and testes, as well as the stress hormones produced by the adrenal glands. It is these downstream gonadal steroid hormones which develop the body's secondary sexual characteristics at puberty and also provide feedback to the hypothalamus for regulation of sexual behaviour. In this way, harmony between the body's basic needs and the brain's ability to address them is synchronously sustained. Synchronisation between the brain and gonadal hormones in the control of human sexual behaviour is not, however, independent of the social environment. A delicate balance persists between those hormones induced by social stress and those which are produced in the context of sexual maturation. The intervening variable of anxiety or fear can be disruptive to the hormonal harmony that regulates sexual behaviour, a harmony that is disturbed by the release of the interfering stress hormones (corticosteroids).

One major difference between human reproduction and that of other mammals is the extended developmental period between birth and puberty onset. In the mouse this period takes place at about 5 weeks of age, while in the monkey this takes place around 2–3 years of age, and in humans, at the start of the twentieth century, puberty commenced around age 16–17 years. In humans, this exceptionally long interval between birth and the onset of reproductive age enables the long process of human brain maturation to become successfully established. Thereby, the executive brain is better able to manage social relationships and to regulate the brain's control over the emotions. Importantly, it is primarily through the knowledge of self, which is gained by social experience, that humans are best able to develop high self-esteem. Thus the extended period from infancy to puberty, longer for humans than for any other primate, has enabled a broadening of secure social relationships. These, in turn, have provided the necessary buffer for ameliorating the emotional turmoil generated by puberty.

When we consider the biology of puberty, it fits nicely into the evolutionary framework of ensuring mammalian reproductive success, and may be investigated at the mechanistic level by experimental animal studies. Females are rate-limiting for reproductive success due to the amount of time and energy they put into their pregnancy, together with undertaking the birth process of parturition, and the postnatal period of nurturing and infant care. The timing of their puberty onset is thus dependent on their body's ability to successfully sustain a pregnancy, and their brain's ability to be sufficiently mature for the successful caring of offspring. This involves a postpubertal lifetime of maternal commitment. The male's biological investment of both time and energy is considerably less than that of the female, namely an overnight production of sperm, the joy of sex repeated at regular intervals and, in the vast majority of mammals, very little in the way of paternal care. It should, therefore, be of no surprise that, in order for the timing of puberty onset and the start of reproduction to be successful, the female should take a lead at the mechanistic level.

KISSPEPTIN: A KEY NEUROPEPTIDE FOR DETERMINING MAMMALIAN PUBERTY ONSET

It is important for female well-being that reproduction should not commence until the body has sufficient reserves of body fat to successfully sustain both the in-utero development of the foetus and its postnatal nurturing. The link between body fat and the effect this has on the timing of puberty onset is complicated, and involves the actions of hormones that engage the hypothalamic–pituitary–gonadal (HPG) axis. Body fat produces the hormone leptin, which acts on the hypothalamus to release the neuropeptide kisspeptin (Herbison, 2016). Kisspeptin is a peptide neurotransmitter which serves as part of the hypothalamic neuroendocrine transduction cascade that underpins the brain's main regulator of reproduction (Klentrou & Pyley, 2003). This it achieves by direct activation of pulsatility in those hypothalamic neurons that produce gonadotrophin-releasing hormones (GnRH). The language of neurons is pulsatile in order for their chemical neurotransmitters not to saturate and thereby down-regulate their target receptor.

Kisspeptin is probably best conceptualised as the gatekeeper for puberty onset (Uenoyama et al., 2016). It is this key neuropeptide which serves to regulate the control of GnRH secretion, which in turn controls the pituitary gland (Figure 4.1). The pituitary gland thus interfaces between the brain and the body's gonads for ultimately determining gonadal hormone production. This neuroendocrine pathway is integral to the onset of puberty. Upregulation of the gain for kisspeptin hormones induces precocious puberty, while downregulation of kisspeptin production delays puberty onset. Genetic factors can also be a cause of precocious pregnancy. A missence mutation in the imprinted gene *MKRN3* has been shown to produce precocious puberty in both females and males (Kansakoski et al., 2015). Mammals have two populations of kisspeptin neurons that are located in their hypothalamus and these have distinct roles in male and female reproduction (Leon et al., 2016). In the female, Kiss1 neuropeptide expression is positively upregulated by oestrogen, whereas in the male, Kiss2 peptidergic neurons are the target for negative regulation by oestrogen. The distribution of these kisspeptin neurons in the hypothalamus is sexually dimorphic (Clarkson & Herbison, 2006), a dimorphism that is thought to be responsible for the sexually differentiated surge of hormones that occur at ovulation and is specific to females. Such differing sensitivities to hormonal feedback may also explain why females are more prone to experience precocious onset of puberty. In part, this is due to the female's increasing levels of fatty adipose tissue in the body, and its production of the hormone leptin.

As well as the potential for body fat to induce early puberty, there are also rare examples of gene mutations that induce early puberty in females. Whereas the body takes the lead for providing the high fat levels that induce early puberty, genetic mutations in the brain may also be the pathological cause of precocious puberty. Hence, mechanisms for puberty onset require a balanced interplay between brain and body. Interestingly, an important gene for activating the onset of puberty onset is the maternally imprinted gene, *MKRN3*, which is part of the same imprinted gene cluster involved in the human Prader–Willi syndrome. Prader–Willi syndrome is characterised by short stature, incomplete sexual development, and chronic hunger that can lead to excessive eating and obesity.

EARLY ONSET OF PUBERTY AND ITS CONSEQUENCES

In the past 100 years, a mere blink of the eye on the human evolutionary timescale, there has been a significant trend towards earlier puberty, especially in young girls. The age at which females normally experience their first menstrual cycles provides an overt sign that ovulation and fertility have commenced, and it is these data which have provided the early documentary evidence as to the age for puberty onset. This kind of information was available long before any measurement, or indeed knowledge of reproductive hormones could be assessed. From the early records collected more than a century ago in Norway and Sweden, the average onset time for females to experience their first menstruation occurred between 16 and 17 years of age. This is considerably later than the present 12–13 years of age for fertile cycles, which now commonly occurs in females found in modern-day Western populations. Healthcare and socioeconomic living standards have undoubtedly improved during recent decades, but of particular importance for the timing of puberty onset has been the availability of high-calorie processed foods.

Body weight, and most notably the amount of body fat, is signalled to the brain via the hormone leptin. Leptin is the critical hormone produced by the adipose fat cells of the body, and it is this hormone which regulates the timing of puberty onset across many mammalian species. Evidence in support of this viewpoint in modern-day humans is abundant. Thus malnutrition is associated with a delay for puberty onset, while moderately obese girls experience earlier menarche than lean girls. Amenorrhoea, characterised by the absence of menstrual cycles, is common in ballerinas who tend to be at the very low end of the scale for fatness, while their body weight is within the normal range. This is also the case for those female athletes who actually have a greater than normal lean body muscle mass, but almost undetectable levels of fat (Klentrou & Pyley, 2003). Hence, it would appear that for human females in particular, the body size and, importantly, sufficient fat reserves to deal with the energetic demands of a potential pregnancy are the important signal for initiating the brain's neuroendocrine cascade that determines puberty onset and fertility. Leptin is a critical permissive hormone in the context of puberty onset, but the neurons expressing kisspeptin are the essential regulators of the GnRH for activating puberty onset (Plant, 2006).

As modern human populations have gained more control over their environment, and thereby assured themselves of a dependable and sustainable availability of food, so too has their timing for puberty initiation commenced earlier. It is the phylogenetically ancient, but biologically efficient, basal parts of the emotional forebrain (hypothalamus) that are responsible for the onset of reproductive fertility, and which also instruct the body to go ahead and breed when nutrition is plentiful. At what is now a very early adolescent age in modern-day humans, the developing executive cortical brain has not matured sufficiently to rationalise control over motivation for sexual activity. Logical and rational reasoning is not, at this early stage, able to apply sufficient braking power to the behaviour of adolescents. This is possibly why modern Western societies have introduced social constraints and cultural laws directed at restraining sexual activity at an early age. Preventing pregnancy and the early arrival of children in advance of the executive brain having undertaken maturational reorganisation is clearly of social importance. Thankfully, the potential social and reproductive problems that can arise from our nutritionally over-supplied fatty diets have, in Western societies, been circumvented by the introduction of efficient contraception.

REORGANISATIONAL CHANGES IN THE BRAIN AT PUBERTY

Longitudinal studies of human brain development from childhood to puberty and onwards through adolescence have shown that the volume of the brain's white matter (fatty myelinated nerve fibres that interconnect multiple regions of the brain) increases linearly from age 4 until 22 years of life (Sowell *et al.*, 2001). However, changes in the volume of the brain's cortical grey matter (neural cells which process the information from the interconnections) is neither linear nor regionally specific. It is within the frontal lobes of the executive brain that the number of grey matter nerve cells increase to their maximum. Increases in these cortical neural cells start at puberty onset, but then undergo a decline in the post-adolescent period due to programmed cell death for those neurons that have failed to consolidate their appropriate connections. Because consolidation of these neural connections is activity-dependent, these neural connection strengths will reflect their owner's social and environmental circumstances. Considering specific cortical brain regions, development of the temporal lobe neuronal cells is nonlinear and peaks at the later post-pubertal age of 16.5 years.

In contrast to our ancestral non-human primates, development of the human executive neocortex is non-synchronous. There are two phases for human cortical brain development at puberty: the early adolescent phase of increase, and the late-adolescent phase of decrease, especially for prefrontal cortical neurons. This has been revealed both by positron emission tomography (PET) studies, and by electroencephalographic (EEG) recordings of brain activity. Such neocortical recordings have discovered a wave of neural proliferation which occurs in the frontal lobes during adolescence, and results in over-production of frontal cortex neurons. This heralds the next stage of post-pubertal development when psychosocial and emotional brain activity guide selective elimination of connections between some neurons, while strengthening the connections of others. Hence during adolescence these regions of the cortex are in a dynamic state of both gaining and losing neurons and their connections. This dynamic state is regulated primarily by epigenetic processes depending on events in the outside world that generate activity in these neurons. Such activity underpins the process of selective connectional reorganisation of neurons, and this in turn programmes the executive regions of the brain for learning to cope with the issues of puberty and adolescence. In-vivo structural imaging studies (high-resolution MRI) from childhood and throughout the period of modern-day adolescence (age 12–16 years) have shown there to be an inverse relationship between the reduction of neurons, but increases in brain size. This is particularly notable for those regions of the frontal cortex that control executive cognitive functioning (Giedd *et al.*, 1999). Such paradoxical findings are thought to represent the loss of those neurons which fail to make appropriate connections (synaptic pruning) together with the progressive myelination of those neurons that do make appropriate connections. Myelination results in neural connections developing a fatty insulating sheath, thereby increasing brain volume and neural communication efficiency within and across the brain. Such neocortical remodelling is also associated with context-dependent improved memory functioning.

SEX DIFFERENCES IN THE REORGANISATIONAL CHANGES OF THE BRAIN AT PUBERTY

Differences in brain structure that synchronise with changing body structure take place in males and females at puberty. These differences are brought about by the male sex hormone testosterone and the female hormone oestrogen (Herting et al., 2014). The human brain's neocortex is extremely complex and few people possess an extensive knowledge of its regional structures, let alone the language that describes these structures. As a simple starting point, we can grossly divide the brain into grey matter (the cellular neurons of the brain) and white matter (the white myelinated axons of the brain which interconnect different component structures, thereby enabling neurons to interact and network with multiple other neurons). In both boys and girls, grey matter volume progressively decreases over the early adolescent period (age 12–16) and white matter volume increases. This has been interpreted as revealing a greater specificity of neural interconnectivity, together with the formation of functional neural networks (white matter increases). Grey matter decreases result from those neurons which fail to make the right connections at the right time, and have thereby not received the activity signals which prevent them from undertaking programmed cell death.

Sex differences are found for the brain's executive neocortical interconnection strengths, particularly those interconnections that provide for cortical asymmetry in the brain. This asymmetry provides females with higher efficiency than males in brain performance for five left hemispheric regions and for one right hemispheric region. Males have greater hemispheric efficiency in two regions of the right hemisphere. The left hemisphere is generally dominant for verbal proficiency and the right hemisphere dominates for spatial abilities. Hence, this asymmetry seen in regional brain efficiency at puberty may underpin the female's advantage in verbal processing while males are more proficient in spatial processing of their environment. In addition to this hemispheric asymmetry, there are other regional differences in the brain that also become reorganised during the adolescent period of males and females. Brain reorganisation during this period is particularly notable in the frontal cortex (an area of the executive brain concerned with cognition, memory and affective functioning) and in the superior temporal cortex (an area of the brain concerned with language), especially on the left side of the brain (Gong *et al.*, 2009).

The frontal lobe's interconnectivities provide a complexity of organisation greater than most other regions of the human brain, and equip us with the ability to thrive in a complex social environment. Decision making, planning and creativity are all part of the frontal cortex's accomplishments. It is this area of the brain which shows changes related to testosterone (the male sex hormone), resulting in the surface area of the male frontal cortex decreasing in size at puberty. In females, this same area of the brain increases in size at puberty. The right superior temporal cortex is also a region of the brain which undergoes reorganisational decreases in size during puberty, and again, these are greater in males than in females. The deep structures within this temporal region of the brain include the amygdala and ventral striatum, which functionally interface with the overlying cortex during 'emotional' brain activity, particularly in the context of 'reward'. This is an area of the brain that becomes larger in boys at puberty (Herting *et al.*, 2015) and may account for adolescent males being more persistent in their behavioural profile, and also more prone to taking addictive substances than females.

Our ability to distinguish between simple facial expressions of emotion also develops over the period of adolescence, and most notably in the female. Females become particularly proficient in their ability to recognise the facial expressions of children. Females also surpass males in their ability to distinguish the real versus false emotions, as well as their ability to emotionally empathise with children (Lawrence *et al.*, 2015). Of course, there is certainly a strong element of learning associated with these positive emotions, as for example, there is for the negative emotions associated with blame, shame and guilt. Here the emotional sensation has remarkable similarities across blame, shame and guilt at the autonomic level. However, the context in which such emotions occur requires the neocortex to distinguish their meaning and thereby develop the relative intensity of the emotional experience. It is almost certainly the case that human emotions experienced in the context of sex and aggression have also been shaped, both positively and negatively, by the use of language and are underpinned by the development of moral and religious beliefs. Maternal emotions for children are incredibly robust, but here too the mother's bonding with her infant may be negatively shaped by the emotions generated through guilt. There is no doubt that there can be a price to pay for our exceptionally language-proficient executive brain. This can elevate the complexity of human emotions to a level higher than in most mammals, primarily through the use of language, thereby distinguishing their meaningful impact and not just via the 'gut emotional feelings'.

Differences in neocortical thickness, surface area and cortical folding at puberty have been shown to occur in males with conduct disorders, especially those males exhibiting callous, unemotional states. Normal adolescent males, relative to females, also show reduced thickness of the superior temporal cortex, and increased folding of these temporal cortical regions together with those of the frontal cortex (Fairchild *et al.*, 2015). Using the very latest imaging technologies, it has been possible to visualise the functional interconnections within these various brain regions and to assess how they transmit information both locally and globally within and across the brain. Females show greater overall cortical interconnectivity, and the underlying organisation of these cortical networks have been shown to function more efficiently (Gong *et al.*, 2009). Connectivity between the two cerebral hemispheres is also found to be greater in women than men. Evidence for this is seen in the larger female corpus callosum, the brain's crossover region for neurons interconnecting the two hemispheres of the cortex. This probably accounts for the more bilateral patterning of brain activation found in women during language acquisition. MRI studies have also found a stronger association between cognitive performance and working memory in women, suggesting their brain can make more proficient use of working memory.

Puberty onset and adolescence thus appear to be very important periods in the process of maturational reprogramming of the human brain's neural connection strengths. This is not just for those parts of the phylogenetically ancient emotional brain regions that are hormonally governed for the motivated behaviours of sex, aggression and maternal care. Puberty also incorporates a period in humans when neural interconnectivity changes occur that enhance more proficient executive brain control. These neural interconnections have not only enabled greater executive neocortical control over limbic brain emotions, but also endow humans with the ability to prioritise certain of our behavioural responses.

The neural density of the frontal cortex reaches a maximum one year earlier in females than in males, corresponding to but not causative for their earlier age of puberty onset. Imaging studies have revealed a significant sex-by-age interaction for the volume of cerebral grey and white matter, suggesting that there are indeed age-related sex differences in the brain's maturational process. In tests of emotional cognition, where subjects are required to make a decision about the emotion expressed in a face, or in a word or both, the reaction times slow significantly at the onset of puberty. Girls start to show this slower reaction time a year in advance of boys, with their reaction times becoming progressively longer between the ages of 15 and 17 years. However, this sex difference is transient and is not present in 18- to 22-year-olds. Such findings are thought to represent a significant difference for the 'age of thoughtful enlightenment' between girls and boys. However, because this age-dependent sex difference also embraces the onset of female fertility, slower reaction times are more likely to be a reflection of the female's more cautious ability in reaction to, and greater rationalisation over, emotive behaviour. Thus it is noteworthy that female adults, compared with adolescents, are better able to engage the frontal cortex (executive brain) when required to attend to different facial expressions, while female adolescents exhibit greater modulation of their emotional brain circuitry based on this same task (Lawrence *et al.*, 2015). This would suggest that the processing of emotional events becomes progressively dominated by the stronger regulatory control by the executive brain, especially in females during the period of their adolescent progression to maturity.

The so-called 'emotional circuitry' of the brain is rich in endorphin receptors. These opioid receptors underpin a very important role in producing feelings of pleasure and well-being, and the contexts for which these receptor neurons are biologically activated adopt strong rewarding characteristics. Many drugs of abuse also act on these opioid receptors, inducing strong states of false 'well-being' that may result in these drugs becoming very addictive (Everitt et al., 2001). Of course, the brain's opioid receptors did not evolve to shape addiction, and the biological context in which they are particularly activated in normal life is that of rewarding actions related to human relationships. None more so than those relationships between the adolescent male and female; teenage 'infatuation', as some parents may call it. However, this is also a period when certain actions can become over-rewarded, and thereby result in the development of obsessive compulsive behaviours, such as the repetitive washing of hands, or the avoidance of lines on the pavement. Considering that the adolescent period is an age for potentially having the first child, then perhaps a degree of obsessive behaviour might also be a real advantage for successful parenting.

Opioid receptors in the brain are biologically important for rewarding social relationships, particularly those of mother and infant. Experimental studies have shown that the effects of these endogenous opioids on the attachment of infant monkeys to their mothers are the same for infants of both sexes early in life. They each attempt to increase their contact time with the mother when their endogenous opioid system is pharmacologically shut down (Martel et al., 1993). At the onset of puberty, however, feral male monkeys in their natural social environment tend to leave their natal social group, whereas the females remain within their social group. This type of social organisation is especially notable in the large-brained Old World monkeys, where it is the females that are seen to remain within their social group. Such group consolidation is referred to as constituting a 'female-bonded' society. Why do we find this sex difference in social bonding, and are these same endogenous opioid mechanisms, which serve for infant attachment with mother, also extended to the context of social bonding? If so, then what is the mechanism for this difference among males compared with females? When males reach puberty, experimental administration of opioid receptor blockade results in them spending significantly more time in proximity to their mother, and decreases the amount of time they interact with other males in the social group. At this same age, adolescent females tend to spend more time alone than do their male counterparts, but when subjected to opioid receptor blockade they increase the time they interact with other females in the group, and not just with their mother (Martel et al., 1995). This illustrates the strong predisposition of females to develop other socially meaningful intrasexual relationships, and thereby expand their secure base beyond the mother. Even so, these additional social relationships are invariably with matrilineal kin. Early post-pubertal males, when under stress, also tend to run to their mother, but this is not favourably tolerated by higher-ranking males in their social group. The adolescent male's desire to be with his mother inevitably invites aggression from the higher-ranking dominant males in the group. Such aggression is possibly a consequence of misinterpretation of the young pubertal male's sexual motives, but the overall consequence results in the exclusion of juvenile males to the periphery of their social group. In this way hormones of puberty also impact, albeit indirectly, on the social organisation of large-brained primates, resulting in the mobility of many young males away from their natal social group.

The attachment that infant monkeys develop with their mother is especially enduring for female offspring and often lasts a lifetime, whereas among males such attachment, inhibited by the presence of dominant males, rarely lasts beyond puberty. These findings also translate to the social organisation of ancient human societies, where it was the young males who likewise tended to be more exploratory, and depart from their social origins.

PUBERTY AND BEHAVIOURAL DISORDERS

The developmental period from puberty into adolescence is a vulnerable time for the onset of many behavioural problems in humans, including eating disorders, obsessive compulsive disorders, addictive disorders and the onset of depression, which in some cases may lead to suicide. Puberty is indeed a high-risk period for suicide. What, then, is taking place in the brain at puberty that makes this a developmental period of high risk for psychiatric conditions, and why do these problems appear to be occurring both more frequently and at a progressively younger age than in previous generations (Frombonne, 1998a)?

In early and middle adolescence there is a marked increase in the symptoms of depression, and this is more prevalent in females than in males (Grant & Compass, 1995). Increased instability and higher demands, both by society and within families, have been proposed to play important etiological roles in depression. This might also explain the increased prevalence of depressive conditions observed in recent decades, especially among young people (Goodyer, 2015). Chronic daily stress and daily aggravation have been strongly associated with depressive symptoms among adolescents, with girls reporting more of these interpersonal stressful events than boys. An extensive study of psychosocial events that lead to depressive symptoms in 12- to 14-year-old adolescents, has revealed strong sociobehavioural correlates. Daily hassles (e.g. being criticised by someone), stressful life events at school in particular, gender issues and a lack of friends all correlated most significantly with depressive symptoms (Sund et al., 2003).

Introspective self-knowledge and self-esteem are central issues in the psychological development of adolescents, and the number of really close friends (four or more) does seem to be protective against the development of adolescent depressive symptoms. Several studies in recent years have revealed an earlier age for the onset of psychosocial disorders, with most of these disorders starting during the early adolescent years (Joyce et al., 1990). Findings over the past three decades have also revealed a progressive overall increase in adolescent psychiatric disorders, and today these disorders are now appearing at a progressively earlier age of onset. It is paradoxical that, at a time when economic conditions and physical health are improving, psychosocial disorders of youth are on the increase, especially in the most financially secure and socially established communities (Frombonne, 1998b).

Schizophrenia is a brain disorder that typically becomes manifest around the time of puberty. Neuroimaging studies point to a progressive reduction in grey matter volume caused by the survival of fewer neurons, especially in the prefrontal cortex of schizophrenic patients. Early puberty in females is associated with later onset of schizophrenia while males show the opposite trend. There is also a notable change in the grey and white matter of the prefrontal cortex at puberty in the brains of schizophrenic patients. Reduction in regional cerebral blood flow, an indirect measure of neural activity, is a further feature of schizophrenia, especially when subjects undertake behavioural tasks that demand activation of their prefrontal cortex. These same patients show neural impairments that are clinically referred to as 'hypofrontal'. In monozygotic twins that are discordant for schizophrenia, the affected twin consistently shows less prefrontal cortical activity in tasks that normally call these regions into action. The link between prefrontal hypofunction and schizophrenia may also involve the temporal lobes of the brain. Such pathologies of the temporal lobe invariably underpin communication through neural pathways that link with the frontal cortical areas. What is clearly manifest from these clinical findings is that schizophrenia, and possibly other brain disorders (bipolar depression), have a strong association with puberty, a period which also reveals notable male/female differences in brain development.

A considerable number of clinical studies have evaluated the possible impact of early, and ongoing, psychosocial stress for the onset and recurrence of unipolar and bipolar affective disorders. In a large population of patients, who had been monitored prospectively with clinical assessments, it was found that those adolescents with a positive history of early adversity (physical or sexual abuse) not only experienced an earlier onset of bipolar illness and more suicide attempts, but these patients were relatively more resistant to treatment in the prospective follow-up (Leverich et al., 2002). Puberty has always been a life event that produces a challenge for well-being, but why is this becoming more of a problem? Can we attribute these increases in psychiatric disorders to failures resulting from early pubertal development, or does the protracted developmental period for neocortical development produce problems in its own right that are exacerbated by modern lifestyles?

Some of these issues have been addressed in recent years by the ability to functionally image the living human brain, a procedure made possible by the advancement of new technologies. Not only can these advanced technologies map regional activity changes in the brain by functional magnetic resonance imaging (fMRI), but they can now determine the interconnective developmental strengths between different regions of the brain by diffusion-weighted imaging (DWI). Finally, transcranial direct current stimulation (tDCS) can assess the relative efficiency of different connection strengths within specific brain regions. Using these new imaging technologies, it has been shown that the executive prefrontal cortex is engaged when humans purposefully assess and appraise the emotional content of facial expressions. Specific regions of this same frontal cortex are actively engaged with the cognitive control over emotions. fMRI studies have further revealed sex differences in the activity of these cortical brain regions during the processing of facial emotional content (Tseng et al., 2016). In females the prefrontal cortex engages initially with the emotional brain (amygdala in particular) in these contexts, while males have a more evaluative neocortical brain response, especially when processing negative emotions. This may explain why females appear able to have a deeper understanding and empathy with the emotions of children, and why males are predisposed to respond aggressively to conflict situations.

When assessing how attractive we find the face of strangers, again, it is the brain's executive prefrontal cortex that is called into action. Studies have shown that with increasing brain excitability brought about by applying tDCS to the right prefrontal cortex, but not to this region on the left side of the brain, there is an increase in the perception of attractiveness for the same face. However, this finding is independent of sex, again suggesting this type of stimulation is probably activating the deeper ventral striatal 'reward' regions of the executive brain. The outcome of these studies failing to differ with respect to either the sex of the viewer or to the attractive evaluation of the facial image would tend to support this interpretation for engagement of the ventral striatal brain reward regions.

Creative cognition has no specific unitary localisation in the brain, but emerges from a complex network of interacting brain regions, and these interacting networks have been shown to differ between males and females when studied by using DWI. Females were found to employ more cortical regions of the brain when producing novel ideas for the resolution of complex tasks. In contrast, males exhibit fewer and weaker positive relationships across the interconnecting networks of these same brain regions when undertaking these same tasks. MRI studies have further shown that adult females have developed higher levels of cortical neuron density and higher functional intraconnectivity than males in these same regions of the brain. Adult male brains have a more distributed organisation of their neocortical networks than do the brains of adult females.

Development of the brain's neural connectivity strengths have been further assessed for males and females prior to and beyond puberty using these same advanced imaging techniques. During childhood the posterior cortical–cerebellar brain connections are primarily to subcortical brain regions and persist into adulthood. This is to be expected because these subcortical neural connections are necessary for walking, eye movement and balance; functions which are fundamental to both children and adults. However, during adolescence the microstructure of some executive brain regions and their neural connections (particularly frontal–cortical and frontal–subcortical) attain levels that are eventually seen in adults. Later to reach maturity are those executive frontal–cortical brain regions which connect with the limbic emotional brain and with the ventral striatum. The latter represent deeper areas of the brain which provide the reward for the learning of those activities which represent the complex interaction between cognition and emotion. These connections take place at mid-adolescence in females, and later, at early adulthood, in males. Such earlier activity dependent neural growth in females is associated with faster, more effective responding and better inhibitory control. In males the later development of white matter (myelination), and hence the maturation of the complex interconnections in the male brain, probably explains why males are less effective in their cognitive control over emotions.

There is no question or doubt as to the complexity involved in development of the human brain, a complexity that is sustained over two decades, especially involving the periods of puberty onset, and throughout the duration of adolescence. These represent the period when neural interconnections are reorganised and finely adjusted in strength according to the kind of environment in which this neural development occurs. Brain development engages multiple epigenetic processes, shaping its organisational connection strengths according to the kind of social interactions which the environment provides. In the context of creativity the brain can be conceptualised as shaping its connection strengths according to the internal thought processes that the brain itself provides.

During development the brain produces an order of magnitude more neurons, approximately threefold more, than the number it eventually retains. The activity generated by these retained neurons is responsible for the making and refining of these neural connection strengths. A scenario of 'use it' or 'lose it' appears to take place, with those neurons that fail to reach their appropriate destination at the right time undertaking 'programmed cell death'. There is no genetic programme that specifies the precision required for the billions of neurons to each extend their axonal connections to the 'right place at the right time'. On the contrary, a genetic programme specifying the trillions of connections (synapses) that need to be made would require a genome too large to be contained in a single cell. Such genetic complexity is rendered relatively simple at the cellular level by incorporation of an 'error elimination' process. Developmental errors are eliminated by the expression of genes that induce neural cell death. It is those neurons that fail to make the right connections at the right time which undergo cell death. This process has been described as representing 'neural Darwinism' by the Nobel Prizewinner Edelman, but neural epigenetics (Waddingtonism) is probably now more appropriate. Such a process of natural selection at the neural level actually results in a brain containing only 33 per cent of all the potentially possible neurons. Should all of these neurons survive, this would certainly present an impossibly chaotic condition for normal brain functioning. Those neurons which do survive are the ones which have been selectively deployed, making the pubertal period particularly vulnerable to adopting inappropriate lifestyles which may shape the subsequent well-being and quality of our future life.

REFERENCES

Apter, D. (2003). The role of leptin in female adolescence. *Ann. N.Y. Acad. Sci.* 997: 64–76.

Blakemore, S. J., Burnett, S. & Dahl, R. E. (2010). The role of puberty in the developing adolescent brain. *Hum. Brain Mapp.* 31: 926–33.

Clarkson, J. & Herbison, A.E. (2006). Postnatal development of kisspeptin neurons in mouse hypothalamus; sexual demorphism and projections to gonadotropin-releasing hormone neurons. *Endocrinology* 147: 5817–25.

Davison, K. K., Susma, E. J. & Birch, L. L. (2003). Percent body fat at age 5 predicts earlier pubertal development among girls at age 9. *Pediatrics* 111: 815–21.

Everitt, B. J., Dickinson, A. & Robbins, T. W. (2001). The neuropsychological basis of addictive behaviour. *Brain Res. Rev.* 36: 129–38.

Fairchild, G., Toschi, N., Hagan, C. C., et al. (2015). Cortical thickness, surface area, and folding alterations in male youths with conduct disorders and varying levels of callous-unemotional traits. *Neuroimage Clin.* 30: 253–60.

Frombonne, E. (1998a). Increased rates of psychosocial disorders in youth. *Eur. Arch. Psychiat. Clin. Neurosci.* 248: 14–21.

(1998b). The management of depression in children and adolescents. In: S. Checkley (ed.), *The Management of Depression*. London: Blackwell, pp. 12–26.

Giedd, J. N., Blumenthal, J., Jeffried, N. O., et al. (1999). Brain development during childhood and adolescence: a longitudinal MRI study. *Nat. Neurosci.* 2: 861–63.

Gong, G., Rosa-Neto, P., Carbonelli, F., et al. (2009). Age- and gender-related differences in the cortical anatomical network. *J. Neurosci.* 29: 15684–93.

Goodyer, I. M. (2015). Genes, environments and depressions in young children. *Arc. Dis. Child.* 100: 1064–69.

Grant, K. E. & Compass, B. E. (1995). Stress and anxious-depressed symtpoms among adolescents: searching for mechanisms of risk. *J. Consult. Clin. Psychol.* 63: 1015–21.

Herbison, A. E. (2016). Control of puberty onset and fertility by gonadotropin-releasing hormone neurons. *Nat. Rev. Encocrinol.* 12: 452–66.

Herting, M. M., Gautam, P., Spielberg, J. M., et al. (2014). The role of testosterone and estradiol in brain volume changes across adolescence: a longitudinal structural MRI study. *Hum. Brain Mapp.* 35: 5633–45.

et al. (2015). A longitudinal study: changes in cortica thickness and surface area during pubertal maturation. *PLos ONE* 10: e0119774.

Joyce, P. R., Oakley-Brown, M. A., Wells, J. E., et al. (1990). Birth cohort trends in major depression: increasing rates and earlier onset in New Zealand. *J. Affect. Disord.* 18: 83–89.

Kansakoski, J., Raivio, T. & Tommiska, J. (2015). A missense mutation in *MKRN3* in a Danish girl with central precocious puberty and her brother with early puberty. *Padiatr. Res.* 78: 709–11.

Keverne, E. B. (1985). Reproductive behaviour. In: C. R. Austin & R. V. Short (eds.), *Reproduction in Mammals: 4. Reproductive Fitness*. 2nd edn. Cambridge: Cambridge University Press, pp. 133–75.

Klentrou, P. & Pyley, M. (2003). Onset of puberty, menstrual frequency, and body fat in elite rhythmic gymnasts compared with normal controls. *Br. J. Med.* 37: 490–94.

Lawrence, K., Campbell, R. & Skuse, D. (2015). Age, gender, and puberty influence the development of facial emotion recognition. *Front. Psychol.* 16: 761.

Leon, S., Fernadois, D., Sull, A., et al. (2016). Beyond the brain – peripheral kisspeptin signaling is essential for promoting endometrial gland development and function. *Sci. Rep.* 6: 29073.

Leverich, G. S., Perez, S., Luckenbaugh, D. A., et al. (2002). Early psychosocial stressors: relationships to suicidality. *Clin. Neurosci. Res.* 2: 161–70.

Martel, F. L., Nevison, C. M., Rayment, F. D., et al. (1993). Opioid receptor blockade reduces maternal affect and social grooming in rhesus monkeys. *Psychoneuroendocrinology* 18: 307–21.

Martel, F. L., Nevison, C. M., Simpson, M. D. A., et al. (1995). Effects of opioid receptor blockade on the social behaviour of rhesus monkeys living in large family groups. *Psychobiology* 28: 71–84.

Mead, M. (1928). *Coming of Age in Samoa.* New York, NY: William Morrow & Co.

Plant, T. M. (2006). The role of KiSS-1 in the regulation of puberty in higher primates. *Eur. J. Endocrinol.* 155: S11–S16.

Sowell, E. R., Thompson, P. M., Tessner, K. D., et al. (2001). Mapping continued brain growth and gray matter density reduction in dorsal frontal cortex: inverse relationships during postadolescent brain maturation. *J. Neurosci.* 21: 8819–29.

Sund, A.-M., Larsson, B. & Wichstrom, L. (2003). Psychosocial correlates of depressive symptoms among 12–14 year-old Norwegian adolescents. *J. Child Psychol. Psychiat.* 44: 588–97.

Tseng, W. L., Thomas, L. A., Harkins, E., et al. (2016). Neural correlates of masked and unmasked face emotion processing in youth and severe mood dysregulation. *Soc. Cogn. Effect. Neurosci.* 11: 78–88.

Uenoyama, Y., Pheng, V., Tsukamura, H., et al. (2016). The roles of kissspeptin revisited: inside and outisde the hypothalamus. *J. Reprod. Dev.* 6: 535–45.

5 Mother–Infant Bonding: The Biological Foundations for Social Life and Cultures

An understanding of the brain mechanisms that underpin mammalian social relationships first requires an understanding of the conserved mechanisms that sustain mammalian maternal care. The mother–infant relationship and maternal care have laid down the foundations upon which the mechanisms for mammalian social organisation were evolutionarily founded. This in turn was brought about primarily through the specialised adaptations arising from mammalian in-utero development and the birth of living offspring. All of the previous vertebrates, from the aquatic fish, the reptilian lizards, the amphibious frogs and the flying birds, have reproduced by laying eggs, and it is within these eggs that the offspring develop. The innovation of in-utero development provided an important niche for many new evolutionary developments including placentation, maintaining a constant body temperature (homeothermy), with a postnatal period of lactation and a commitment to maternal care. These evolutionary specialisations have given rise to significant differences between the brains and indeed the behaviour of males and females, differences that were especially driven through the matriline. Female mammals invest extensive amounts of their time and energy in pregnancy and parenting behaviour compared with males. High levels of prenatal resources are supplied to the developing embryo via the placenta which, together with the unique ability of female mammals to produce postnatal milk, has further devolved the postnatal priority for parental care to the mother. Infants thereby became special to mothers through the engagement of neural 'reward' mechanisms in the mother's brain. The nature of this 'infant reward' is linked to the mother's hormonal state and to the specialised selective mother–infant recognition systems. As many of the early mammals lived underground or built nests in well-camouflaged regions, their primary recognition system was, and still remains, that of olfaction (Keverne, 2002). Large regions of the mammalian genome (> 1000 genes) are given over to coding for olfactory receptors that respond to volatile odours, as well as vomeronasal pheromone receptors (> 350 genes) for non-volatile proteins. These pheromonal receptors are located in a specialised vomeronasal cavity and respond to non-volatile cues, providing most mammals with the specific information needed for recognition of individuals, sexual status, pregnancy and social status. Pheromonal cues require intimate contact with the individual or with their urine and environmental secretory marks. The communicatory importance of these pheromones has been revealed by experimentally producing lesions to the receptor neurons or to deleting the genes that code for pheromone receptors in mice. Such chemosensory-depleted mice suffer major social, reproductive and behavioural deficits including impaired mother–infant care, impaired sexual behaviour especially by males, and a failure to make gender recognition (Stowers *et al.*, 2002). Thus a common biology, linked to olfaction and pheromones, underpins many aspects of the social, sexual and maternal behaviour that are observed among such small-brained mammals. In these mammals, chemosensory-driven behaviour and endocrinology can be considered as integral to their reproductive success.

In considering the reward mechanisms of the brain that are linked to hormonal status, there is one such hormone in particular, namely oxytocin, that has been integral to the evolution of mammalian motherhood (Insel & Young, 2000). Oxytocin is a neuropeptide produced in the brain and also released from the posterior pituitary gland into the bloodstream, where it is important for initiating uterine birth contractions as well as stimulating the mammary glands for milk let-down. In the brain itself, the oxytocin peptide is released directly from neurons, and is important for promoting maternal care and mother–infant bonding. During late pregnancy, the receptors for the oxytocin protein are upregulated in both the brain and the uterus of mother. This occurs in response to hormonal signals released from the foetal placenta, most notably during the transition from the release of the hormone progesterone throughout most of pregnancy, to the production and release of oestrogen during the late stage of pregnancy. In mammals such as rodents and sheep, mother–infant recognition bonding is achieved through the release of oxytocin in the mother's brain, which occurs in response to giving birth. Primed via the action of the placental steroid hormones during pregnancy, the mother's brain develops the receptors for oxytocin, and at parturition her brain is ready to respond to oxytocin release which, in turn, promotes maternal care (Curley & Keverne, 2005). The birth process itself, parturition, is common to infants of both sexes, as both possess a placenta and both benefit from maternal care. It is the placental hormones, independent of foetal sex, that promote uterine birth contractions and prime the adult female brain for maternal behaviour. However, because adult males never experience pregnancy, this neuroendocrine underpinning of maternal care fails to become engaged in males. Typically, male mammals such as rodents do not form lasting bonds with their offspring, and unlike females, their adult social relationships are characterised more by aggressive behaviour than by affiliative behaviour. Indeed, most of the male species in possession of a small brain rarely stay with their pregnant female and will actually kill newborn infants should they encounter them. A hormone closely related to oxytocin, vasopressin, is made in the same region of the male's hypothalamic brain as is the production of oxytocin in the female brain. It is this hormone, vasopressin, which under the influence of testosterone in males regulates male-typical behaviours such as territorial marking and aggression while suppressing their parental care of infants (Young, 1999).

When we consider those mammals which possess large brains, notably the monkeys, apes and especially humans, there is a decreased reliance on olfactory information for the initiation of most aspects of behaviour, including parenting behaviour. Indeed, the vomeronasal receptor genes have themselves become non-functional pseudogenes in most primates and especially in humans. In humans the genes which code for these vomeronasal chemoreceptors remain as sequences in the genome in the form of 'junk' DNA, and are not available for expression. Redundant DNA sequences are thus present, but gene transcription fails to take place for the pheromone protein receptors. In the human genome some 60 per cent of the main olfactory receptor genes are non-coding pseudogenes (Mori et al., 1985). This feature has, in turn, been accompanied by a reduction in the proportion of the olfactory cortex given over to processing olfactory information (Figure 5.1). The decreased dependence on chemosensory cues across mammalian phylogenies is further reflected in the relative increase in size of their neocortex relative to the olfactory cortex. Such a decline in both types of chemosensory odour processing has been evolutionarily driven by the ability of primates, especially those possessing a large neocortex, to gather their social and foraging information from visual cues as they evolved away from the nocturnal to a diurnal lifestyle. Arguably the most significant change to the visual system of large-brained primates was the evolution of colour vision, which occurred at approximately the same time in evolution as the vomeronasal receptor genes functionally declined. Colour vision enabled primates to interpret social information at a distance. Such information signalling is provided by the colourful sexual adornments around the genitalia that signal reproductive receptivity in females (Dixon et al., 2005). In males the communication of their dominance status is also signalled by visual cues, notably the enlargement of canine teeth, and facial colouration changes, which focus attention to the face and to the mouth in particular. Moreover, sexual behaviour is no longer restricted to periods of reproductive fertility in these large-brained primates. This is especially the case in Old World monkeys, apes and humans. Indeed, most sexual interactions in humans have become non-reproductive but, importantly, these interactions do serve for bonding. Moreover, bonding by mother and infant in these large-brained primates has become an integral

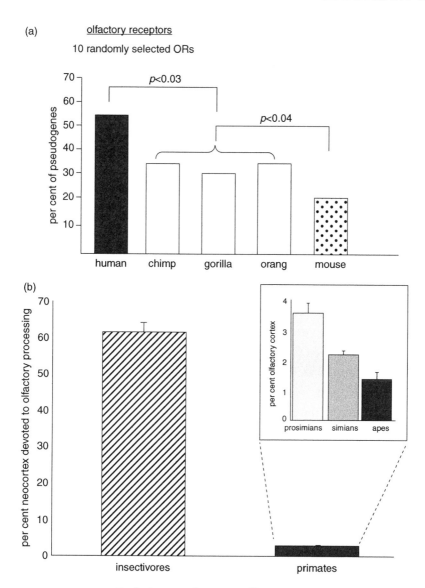

FIGURE 5.1 Evolutionary reduction in olfaction receptor survival and reduction in olfactory cortex. (a) With evolutionary progression from the mouse, through the great apes to humans, many of the olfactory receptor genes progressively become non-functional pseudogenes. (b) In line with this olfactory gene dysfunction there is a proportionate loss of the neocortex that is given over to processing this information.

part of maternal care, extending the period of the mother–infant relationship beyond the biological boundaries set by parturition and even beyond the period of suckling. As infants became progressively more mobile, their ability for recognition becomes more dependent on the visual system, consequently enabling individual facial recognition at a distance.

Those parts of the executive cortex that process visual information have become especially enlarged in primates, and for humans in particular these areas of the brain now comprise up to 50 per cent of the total neocortex (Striedter, 2005). Visual association areas of the cortex have also become increasingly more complex, with several regions of the executive brain being devoted to the differential processing of visual information, most notably facial recognition and facial expressions. Indicative of this evolutionary progression from a reliance on the proximate olfactory cues in small-brained rodents to the more important role for visual information in large-brained primates has been an increase in the size of the nucleus that relays visual information. This lateral geniculate nucleus is central for the relaying to, and the subsequent higher-order processing of, visual information in the neocortex and association cortex. At this same evolutionary time, the mammalian olfactory bulbs which relay their chemosensory information to the ancient paleocortex have become relatively diminished in size, especially with respect to the rest of the brain. Indeed, throughout mammalian evolution, all of the olfactory processing regions of the brain have failed to increase in line with the continued expansion of the executive neocortex. This six-layered neocortex, which has progressed to ever-more complex levels of organisation, is important for the storage and integration of visual information processing. Moreover, visual information, unlike olfactory information, is not neurally directed to the more primitive three-layered allocortex, but is processed by the specialised six-layered visual neocortex. With increased neocortical specialisation for visual processing, primates have become more dependent on complex colour vision, especially for their development of discrete social recognition. Visual recognition has become the primary sensory system for recognising kin and other social alliances around which the foundations of social structure became organised.

THE FORMATION AND REGULATION OF PRIMATE SOCIAL BONDS

The transition from assessing social information based on contact with pheromonal chemical cues to the deployment of visual cues at a distance further witnessed a significant advance in the evolution of a larger brain, and the deployment of complex behavioural gestures in the monkeys and apes. In parallel with brain enlargement, a further major evolutionary advance also took place in the number and complexity of social relationships that could be sustained by the brain. It is this individual recognition and social knowledge of the many relationships generated by group living which have been integral to interpersonal cooperation and stability across and within primate social groups. This in turn has led to the development of social bonding. Interestingly, the mechanisms by which these social bonds are sustained by the brain have their origins in the very same conserved mechanisms that underpin and sustain the mother–infant bond (Broad et al., 2006). Primate societies are recognised for the complexity of their social organisation and for the increased cognitive skills required to process and act upon this social information. Thus, throughout primate evolution, there has been a general trend for increasing brain size (encephalisation) relative to body weight. This is particularly notable for those primate species living in complex social groups, such as is found in rhesus monkeys, baboons and chimpanzees (Curley & Keverne, 2005) (Figure 5.2).

Evolutionary enlargement of the primate brain has not taken place isometrically, and different brain regions have evolved at different rates. It is those parts of the brain that are concerned with higher-order cognitive capacities (neocortex, striatum and hippocampus) and collectively termed the 'executive' brain which have

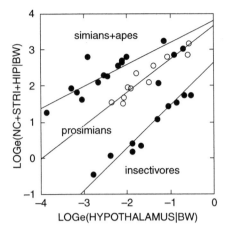

FIGURE 5.2 With growth in group complexity across several groups of the simian monkeys and the apes, progressively more neocortex relative to the hypothalamus has developed. Insectivorous mammals are very low on the logarithmic scale for neocortex, but relative to body weight are very similar for the hypothalamic brain regions. *Abbreviations*: NC, neocortex; STRI, striatum; HIP, hippocampus; BW, brain bodyweight.

shown the most significant evolutionary increases in size across mammalian evolution. These increases in brain size range exponentially from the very small insectivorous foragers to the socially complex primates. Those more ancient parts of the primate brain (hypothalamus, preoptic area and amygdala, which are collectively termed the emotional or 'limbic' brain) are involved in regulating the hormonal control of primary motivated behaviours. These include sexual behaviour, aggression and maternal care, and these regions of the brain show remarkable structural stability across most mammalian lineages. However, relative to the size of the expanded executive brain, these limbic brain regions have become significantly reduced in volume. Thus, whereas the processing of social bonds in small-brained mammals requires a hormonal context and depends primarily on chemosensory information, in large-brained primates there has been considerable evolutionary advances in the neocortex. In these large-brained primates we see the emancipation from the hormonal determinants for all aspects of social behaviour, and this has been replaced by a greater dependence on social learning which is carried out by the executive brain.

There is no denying that hormones are essential in other biological contexts, even for large-brained primates (they are still needed to support the physiology of reproduction and pregnancy), but in a social context, it is essential to emphasise the increased significance for the higher-order cognitive functioning that is undertaken by the executive brain. It is this executive functioning of the brain that has enabled social and sexual interactions to occur outside the restrictive boundaries set by those hormonal determinants seen in small-brained mammals. Moreover, it is this increased growth of the executive brain and its processing power that has provided the crucially important advance in the determination of social status. Social status has become integral to the evolutionary underpinning of primate social organisation, while growth increases in the primate executive brain have, in turn, required longer postnatal care for ensuring this developmental growth of the neocortex. Such care extends well beyond the post-partum period of lactation, and has therefore required the maternal maintenance of this care to take place independently of hormonal control. Emancipation of maternal behaviour from hormonal determinants has thus enabled younger, non-parturient females in the social group to assist with infant care. In turn, the experiences of alloparenting develop and improve the infant-caring skills of these adolescent females. Indeed, many young female monkeys are spontaneously maternal without the need for either pregnancy or for hormonal priming. Young females do, however, need to learn and develop these maternal skills under the watchful eye of a mother (Broad et al., 2006). Laboratory female monkeys that have not experienced the early lifetime opportunity to interact with other infants fail to develop their maternal skills, and are often abusive even towards their own infant. Thus an evolutionary transition from the 'hormonal biological' regulation over decision making to an integrated 'cognitive' control over decision making has been an essential step for the successful outcome of simian primate maternal care and, indeed, for many of primate life's social relationships. These, in turn, have markedly shaped the functional developmental enlargement of the brain, enabling the recognition of subtle social cues and their appropriate contexts. In this way, social learning has become important for determining group stability, and hence a relatively peaceful organisation within the primate group.

The mother–infant relationship for those mammals with small brains is primarily dependent on hormonal activation of the oxytocin neuropeptides contingent on pregnancy and parturition. In contrast, those primates with a large brain sustain their social bonds continuously throughout life by affiliative interactions within the social group. Although not initiated by hormonal priming, primate social relationships do, nevertheless, require a biological 'social glue' acting within the brain, a 'glue' which is necessary for reinforcing affiliation and bonding (Vines, 1996). Consequently, the evolution of additional brain mechanisms for rewarding and sustaining social bonds have been essential to the continuity of social stability within large-brained primate groups. Such social bonds develop under the control of the executive brain's 'reward' system. Thus, in many of the large-brained primate species, their social bonding is achieved and reinforced through the tactile stimulation provided by social grooming and huddling. This social behaviour initiates the release of the executive brains rewarding opioid neuropeptide, β-endorphin (Figure 5.3). Indeed, there is increasing evidence for social grooming acting as a primary stimulus for endogenous opioid release, thereby providing the interactive reward for maintaining their social bonds (Keverne, 2005). Moreover, it has been observed that those female baboons who are groomed more frequently by others (i.e. are more social) have higher rates of infant survival and fitness. In addition, the number of different grooming relationships (an indicator of social status) is one of the most accurate predictors of neocortex size across different primate species. Although β-endorphin also underpins 'reward' in small-brained mammals during mating and parturition, in primates this neuropeptide has acquired the distinctly

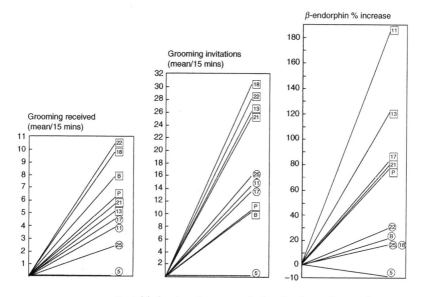

FIGURE 5.3 Social behaviour increases the levels of social rewarding opioid peptides (β-endorphin). Tactile grooming invitations and grooming received represents a powerful social bonding stimulus. Four of the top six monkeys for grooming behaviours show the highest increase in β-endorphin, a social rewarding opioid neuropeptide.

unique function of rewarding social grooming encounters, thereby promoting and sustaining multiple individual closely bonded relationships (Martel et al., 1993). Importantly, the effects of blocking the opioid receptor activates a greater grooming response from the females, as if they are attempting to compensate for opioid withdrawals (Figure 5.4, left). Administration of the opioid receptor agonist (morphine) decreases the social grooming among these females in the primate group (Figure 5.4, right). Such social interactions no longer provide the same reward when the endogenous opioid system is made dysfunctional. Grooming behaviour is also reflected in social status, particularly among females, who are the primary participators in this affiliative interaction. Moreover, it is the females of higher social rank who participate more in this social interaction and also receive most of the grooming in their social group (Figure 5.5). Hence, social grooming, through the release of endorphins, may be considered as

122 BEYOND SEX DIFFERENCES

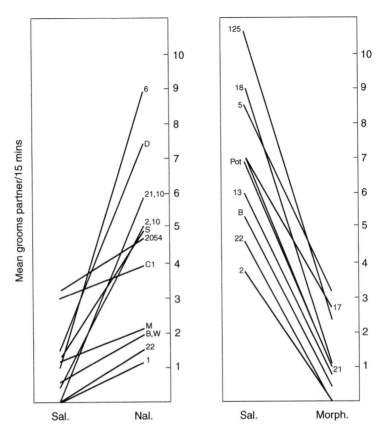

FIGURE 5.4 Effects of the opioid receptor blocker (left) and administration of opioid agonist (right) on social grooming of female partner. Opioid peptides are powerful rewarding neuropeptides for social interactions. The opiate receptor blocker (naloxone) is a powerful stimulus for grooming, i.e. taking away the acute reward for grooming increases the grooming interaction (to compensate). Administering the opioid agonist (morphine) has the opposing effect of decreasing grooming action (the behaviour is no longer sufficiently rewarding).

providing the primary reward for those social interactions that sustain group cohesion.

There are only a few monogamous non-primate mammals that bond for life, and the monogamous prairie vole is one such example. This relatively small-brained mammal has adapted to harsh environments and only survives by male cooperative nest-guarding together

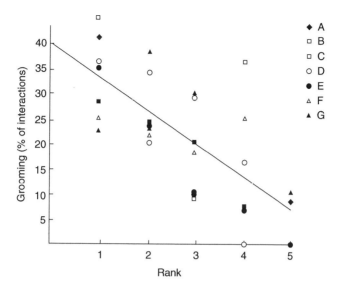

FIGURE 5.5 Monitoring the grooming interactions of monkeys according to rank. Primates form social hierarchies according to their dominance/subordination behaviours. Grooming interactions are correlated with rank, which is determined from the direction of aggressive interactions.

with joint parental care of infants. The neural mechanism for this monogamous type of pair-bonding has some similarities to that which is found in primate social bonding, but is somewhat less complex. Prairie voles are characterised by higher levels of expression for the oxytocin receptor, enabling the oxytocin hormone to act in brain areas associated with reward, but here the signalling of the reward stimulus is primarily recognised by olfaction (Figure 5.6) The genetic basis for this male–female pair-bonding in prairie voles also has mechanisms in common with mother–infant bonding, and has not therefore required to be underpinned by the evolution of novel genes. However, altering the upstream regulatory sequence of the gene which determines where the oxytocin receptor is expressed in the brain has been important. In other words, these voles capitalise on the neural mechanisms that are evolutionarily well established, namely those that are called into action for maternal care in the female and by mating behaviour in the male of small-brained mammals.

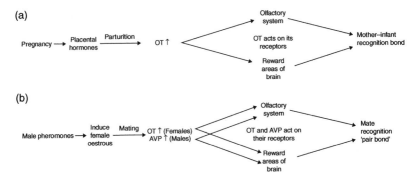

FIGURE 5.6 Bonding in small-brained mammals. Mother–infant bonding and pair-bonding in small-brained mammals have several conserved features. In mother–infant bonding (a), the hormones of pregnancy provide the context. Pair-bonding in the monogamous vole (b) is dependent on mating. Oxytocin provides the 'bonding peptide' for mother and infant bonding in small-brained mammals and is signalled by olfactory cues. Oxytocin and vasopressin provides the stimulus from mating-induced oestrous. Both types of bonding are initiated by odour cues and each depends on the stimulus of vagino-cervical stimulation, mating and parturition.

The ultimate evolutionary selection pressure for monogamy in this minority of small monogamous mammals is thought to have been determined by their sparsely distributed population dynamics within harsh environments. In such environments there is little opportunity for the male to find a new mate, and should he desert the established partnership, then his infants are at risk for survival, thereby resulting in the decline of his reproductive success as well as that of the female partner.

Briefly summing up, the transitions for bonding from mammals with small brains to those primates with larger brains that live in socially complex groups has been based on both neurological and physiological developments. The major functional and anatomical changes to the primate brain reflect this evolution. Most of the monkeys and apes that possess a large executive cortex are not heavily invested in olfactory behaviour, but like humans they are committed to complex visual processing. The transition from a nocturnal to a diurnal lifestyle among arboreal foraging primates has also placed strong selection pressures for their development of colour vision. Living in large groups has further required a 'social glue', namely the rewarding endorphins that have fostered the development of close relationships between mother and infant. These relationships endure beyond the episodic biological life events of mating, or pregnancy and parturition. Such episodic biological life events are both necessary and sufficient for underpinning the transient, hormonally dependent, periodic mothering relationships of small-brained mammals. The complexity of primate social groups has further provided a context for which their social bonding is not simply determined by hormones, but is activated by, and is dependent on, their social interactions. Social status in the group is also determined from the group's complex behavioural interactions, their knowledge of which is dependent on social learning. Moreover, the tactile grooming and huddling that frequently occurs among dyads and triads within the group primarily takes place among members of the extended and socially related matriline. Such tactile social interactions activate the release of those neurotransmitter signals in the brain, dopamine and endorphins, that provide for 'social reward' (Martel et al., 1995) (Figure 5.4). Thus the development of a large brain has, to a great extent, emancipated reproductive decision making from the relatively sporadic biology provided by the body's hormonal secretions. Such hormonal determinants have been replaced by a dependence on the greater complexity of social knowledge that stems from social relationships, especially from those relationships which are sustained via the extended matriline. These social relationships have, in turn, produced important consequences for interactions between the sexes, and for female–infant grooming interactions. Of special importance for infants are these early grooming interactions with their mother, and subsequently with her extended matrilineal relatives. In this way, social bonding fulfils more than just a reproductive role, and mother–infant bonding is neither restricted to the hormones of pregnancy nor limited to the post-partum period, as occurs in those mammals possessing relatively small brains.

Parenting and alloparenting are lifetime occupations for group-living female primates, and this has had a profound impact on their brain evolution. Brain growth is energetically costly, but postponing the greater part of this to the post-partum period has not only conserved the energetic demands on pregnancy, but has also ensured that the brain itself undergoes most of its development in an environment that facilitates a social lifestyle. Social living and the development of a large executive brain have thus become integral to achieving both selective adult bonding and, in turn, a lifetime's reproductive success.

SOCIAL ADVERSITY: INFLUENCE ON BRAIN DEVELOPMENT

An evolutionary increase in development of the neocortex is to be found in both human and non-human primates, and these species are characterised by a dependence on the complex nature of social interactions that are integral to their lifetime survival. At birth, the developing infant's brain receives social stimulation from a mother committed to providing the emotional rewards of suckling, huddling and grooming. It is clear that the process of further socialisation during social group interactions also benefits strongly from the close relationship that is established between infant, mother and the family group. Because the mother–infant relationship develops during the early postnatal period of brain development, then mother–infant separations are likely to have long-term adverse consequences for infant development. Indeed, negative consequences for both their social relationships and maternal bonding during the next generation are found for those infants that have been separated from their mother and reared with peers (Kraemer et al., 1991). Such nursery-reared monkeys, deprived of a maternal upbringing, develop a limited behavioural repertoire and have fewer social interactions later in life. These monkeys may also develop an increase in stereotypical behaviours. Increased stereotypical behaviours are classically associated with the disruption of executive brain function, particularly frontal cortex function, and result from the inability to suppress inappropriate repetitive behavioural responses. Such repetitive behaviour has much in common at the neural mechanistic level with the obsessive compulsive stereotypies seen in psychologically disturbed human children.

Squirrel monkeys, four years after experiencing separation from their mother during early development, suffer consequent impairments to their developing brain, notably involving those regions of the medial prefrontal executive cortex. Behavioural tests have subsequently shown these monkeys to be impaired in those mechanisms that underpin brain social reward (Lyons & Schatzberg, 2003). Electrophysiological recordings from normal adult monkeys have shown that it is the prefrontal cortex and striatum of the executive brain which mediate these mechanisms for reward (Matsumoto & Tanaka, 2004). Such involvement of the prefrontal cortex and ventral striatum in reward-related behaviour has also been observed in humans by the employment of brain imaging techniques. Interestingly, it is the human ability to detect unfavourable outcomes, response conflict and decision uncertainty which activate overlapping clusters of neurons in these same cortical regions of the human brain that signal to the ventral striatum. Choosing between actions associated with uncertain rewards and punishments in humans is also dependent on these same regions of the frontal cortex neural circuitry that signal 'reward'. Moreover, when human mothers watch alternating videos of their own child versus the child of a stranger, the greatest brain signal contrasts occur within, and are restricted to, distinct regions of the frontal cortex. Such refined visual distinctions for recognition of own versus strange infants involve complex components of face recognition together with the emotional processing of the face. Hence, facial recognition not only engages activation in the visual cortex, but the emotional content of the face requires further integrative processing by the temporal cortex and the amygdala (Ranote et al., 2004). Thus, what we may consider to be a relatively simple task actually engages many areas of the human executive as well as the limbic emotional brain during the recognition of infants. It is these very same executive areas of the brain that have become uniquely enlarged in humans and that are also activated during the recognition of different human emotions. Although human emotions have much in common with animal emotions at the autonomic level, the use of language has further extended the human emotional repertoire to a higher cognitive level, a level dependent on social context. A good example of this is found with 'shame, blame and guilt' which, at the autonomic level (sweating, dry mouth, stomach churning) are indistinguishable emotions. However, given the contextual definition provided by language, together with the interpersonal relationships that are involved, this social information can amplify or diminish the meaning and intensity of the emotional feelings experienced. Thus, the development and understanding of language has introduced yet another dimension, namely one which provides a complexity to those mechanisms by which the human brain processes emotional information. Put simplistically, the human brain processes emotional events by expanding and enhancing their 'gestalt', a process which is achieved by employing a combination of human executive brain rationalisation of the emotional brain's limited repertoire of 'gut feelings'.

INFANT–MOTHER ATTACHMENT

A feature of particular consequence for human social relationships goes beyond the unidirectional mother–infant bonding, because the infant is itself considered to play an active role in the development of this earliest of relationships with its mother. This 'attachment' relationship is considered to lay down the foundations for later socioemotional development. The child developmental psychologist, John Bowlby, postulated that experiences within the early relationship between infant and mother serve as an internal working model of 'self' in relationship to others. This working model was proposed to serve for guiding future interpersonal interactions. The security of this attachment with the mother further provides a sound base for the expanding of relationships that are trustworthy with others. Attachment behaviour starts early in life with the survival benefit of protection from harm (Bowlby, 1982). To quote Bowlby, 'the biological strategy of attachment behaviour in young infants has evolved in parallel with the complementary parental strategy of responsive care giving. The one presumes the other'. The attachment behaviour of infants with their mother, and the use of this relationship as a 'secure base', thereby enables infants to confidently explore their own expanding social world. In this way, the pattern of communication that a child adopts towards the mother comes to match the pattern of communication that the mother has been adapting towards the child (Bowlby, 1991). The sensitivity of the mother's responsiveness to her infant is dependent on her accuracy in reading the infant's attention-seeking signals and responding appropriately, thereby providing a 'security of attachment'. The mother's sensitivity, although a reliable promoter of secure attachment, is not, however, independent of her own emotional state (anxiety or depression) and can also be negatively influenced by her own previous traumatic experiences. However, for the most part, infant–mother security of attachment is mainly associated with positive outcomes.

Bowlby, unlike most child psychiatrists, was more concerned with a child's experience of life events, especially when it came to understanding the trajectory for the child's psychological development. He was more focussed on the developmental enrichment of the infant through a close emotional relationship with the mother. However, under normal circumstances the mother integrates her experiences with her child according to those experiences from her own past history of close relationships (Fonagy, 1999). Thus, it is the mother who interprets the behaviour of her child based on her own experiences from when she was growing up with her parents and family. As a consequence, the mother's way of interacting with her child is also thought to be, at least in part, an expression of her own attachment history. On the basis of this model of attachment, the quality of attachment bonds between mother and child are intergenerationally transmittable.

The method of assessing the past history of attachment is mainly by interview, which takes into account and controls for general intelligence and verbal ability. In the case of child assessment, a procedure known as the 'Strange Situation' is used; this is a situation in which an increasing amount of mild stress is placed upon the infant (Ainsworth, 1969). The assessment involves a strange place, an unfamiliar person, and the experience of being alone, followed by the child being reunited with the mother, all of which is videotaped for subsequent analysis. In this situation the behaviour of the mother is purposefully controlled, and only the behaviour of the child is used to assess the child's attachment security. There are three attachment classifications: 'secure', for the child who uses the mother as a secure base for exploring the environment and sharing objects, such as their new toys, with mother. 'Insecure avoidant' attachment refers to a situation when the infant fails to engage in sharing, and shows little 'positive affect' with the mother on reunion. Infants who are classed as 'insecure avoidant' do little exploration in the strange situation, are wary of strangers, continually monitor the mother, and become highly stressed when separated from her.

When it comes to assessing the state of mind for attachment in adults, the 'Adult Assessment Interview' is used. This is structured entirely around the topic of attachment in the context of the individual's relationship to their mother and father (Hesse, 1999). The interview is not designed to test the adult's actual experiences of childhood, but how they currently reconstruct the meaning behind these childhood experiences. A rating scale is applied to these data to assess the way these adults currently evaluate such early experiences. Three main classifications have also been devised for adults. 'Secure autonomous' reflects the mother's evaluation of her own 'attachment'-related experience. Adults that are placed in this category tend to be self-reliant, objective and not defensive in their behaviour. Indeed, they tend to report a history of supportive experiences. The 'Insecure dismissive' classification of adult behaviour tends to deny any negative experiences, and provides an idealised picture of their parents that is inconsistent with other parts of their appraisal. Finally, adults that are classed as 'Insecure preoccupied' are often confused and lack objectivity about their parental relationships, revealing unresolved issues from which they are unable to build a secure dependence.

Some of these common themes have been found across child and adult classifications of attachment, and might therefore suggest that attachment is inter-generationally transmitted. van IJzendoorn and De Wolff (1995) carried out a meta-analysis of attachment research on both adults and children, which they interpreted as providing clear evidence for concordance of parent and child attachment, although this concordance was far from perfect. This is not to deny a heritability component for attachment classifications, but any suggestion that this heritability is genetically transmitted is extremely questionable. Studies in twins (identical versus fraternal) has concluded that security of attachment is not demonstrably genetically heritable. They found only modest 'genetic influences', while a substantial environmental influence was unquestionably present (Bokhorst et al., 2004; O'Connor & Croft, 2001).

There has been a tendency to restrict attachment theory to the earliest stages of child development. Without doubt, this early period is undeniably significant in child development. However, the pathway along which attachment further develops continues to be shaped by offspring experiences throughout their infancy and childhood, and even extending into adolescence (Bowlby, 1980). Attachment experiences initially result in 'internal working models' or 'expectations' which begin from a relatively early age. Moreover, recent studies leave little doubt that there are long-term and lasting influences from childhood attachment experiences, and that these experiences influence both intersocial and personality development. Disruption of infant attachment does sometimes occur as a result of events such as maternal depression, maternal drug addiction and infant health problems at birth. Maternal clinical depression which has taken place prior to infant birth has also been found to predict incoherent and less-sensitive care-giving by the mother. The mother's biological rhythms, if they are disrupted, can also impact on the mother–infant interaction, especially in the context of maternal feeding which may, in turn, produce an interactive refusal to be fed by the infant (Santona et al., 2015).

Maternal drug abuse is unquestionably a major risk factor that can influence maternal functioning at multiple levels, leading to less than optimal positive interactive exchanges between the mother and her child (Porreca et al., 2016). Drug abusers themselves often report negative or traumatic attachment representations from their own childhood. Interestingly, those infants that had not been separated from their addicted mother show better interactive behaviour during nursing. These mothers, despite their addiction, remain sensitive, responsive and more positive in their affect, leading to a more secure relationship than might have been expected with their infant. However, it is clear that in the main, drug addiction in adulthood certainly increases the risk of suboptimal care-giving for the next generation, thereby perpetuating the intergenerational cycle of early adverse experiences and addiction (Kim et al., 2016).

The question arises as to what might be the mechanisms that underpin security of attachment. Studies on same-sex monozygotic or dizygotic twin pairs at three and a half years of age have found no evidence for a genetic contribution to their attachment patterns (O'Connor & Croft, 2001). The level of concordance across the groups for secure/insecure attachment (70 per cent versus 64 per cent) was very similar, and provided no support for any genetic influence. However, if one of the monozygotic identical twins displayed disorganised attachment, the other twin was sometimes similarly rated, while among dizygotic twins there was a zero concordance. It is nonetheless possible, indeed extremely likely, that the observed concordance for monozygotic twins could have been a result of similar shared environments, because these twins are simultaneously reared together by the same mother. Of course, genes do not code for concepts such as attachment, or how they might differ between 'secure' versus 'insecure avoidant' attachment. In the context of behaviour, genes code for the development of neurons. Neuron survival, or neuron programmed cell death, depends upon the synaptic connections that neurons make in the brain and whether these occur at the right time and in the right place. Moreover, the correct formation of these neural interconnections is very much a part of epigenetics, dependent on neural activity, and has much in common with learning and memory. Learning and memory is not in itself sufficient explanation for the context of secure attachment, because the quality of attachment is further dependent on concurrent neural activity in the ventral striatum, the brain's 'reward centre'. Interestingly, maternal deprivation in experimental animals does have long-term effects on the volume, number and size of neurons in this 'reward' region of the brain (Aleksic *et al.*, 2016). Moreover, epigenetic processes in the brain are known to be activated by prosocial mothering in rodents, which in turn produces prosocial mothering for the next generation, even when the mother is not genetically related (Champagne, 2012).

THE HISTORY OF HUMAN SOCIAL EVOLUTION

For centuries humans have collectively engaged in trying to construct environments that provide the kind of physical and emotional security that they subconsciously experienced during their protected in-utero and early postnatal development. In doing so, these physically adaptive environments have ameliorated infant lives from the stresses of climate variation and the anxieties of food uncertainty, thereby advancing the survival trajectory for their own evolution and behaviour. Maintaining these physically secure environments has required transgenerational learning of culturally transmitted traits. This learning does not become encoded in our heritable genes, but is nevertheless capable of transmission across the generations through long-lasting epigenetic changes to the brain which are an integral part of long-term memory formation. Such positive outcomes are reflected in the long-term changes to the neural interconnections that take place in the brain. The downside of such epigenetic malleability of the brain is that poor environments, especially those involving early social deprivation or environmentally chronic stressors, can have equally long-lasting effects which, unfortunately, may produce social and psychological problems in later life (Chungani et al., 2001).

Such is the capacity of the human brain's functional ability for forward planning that our biological needs can be delayed or even ignored. Thus, the human behavioural approach to engage in feeding is rarely determined by hunger; modern-day humans cultivate and harvest food to avoid hunger, but for cultural/religious reasons we may, on occasions, also deny ourselves foods and ignore the hunger. Likewise, most human sexual activity is biologically non-reproductive, and women can become excellent mothers without undergoing exposure to the hormones provided by pregnancy and parturition. In the context of aggression, fighting by our male tribal ancestors had the immediate potential of enhancing their inclusive fitness by the gain of material resources and mates. However, in modern societies the motivations behind large-scale conflicts and international warfare have little impact on the inclusive fitness of those individuals who do the fighting. This is not to be interpreted as illustrating behavioural brain rewards as unimportant, but to illustrate the human capacity for self denial, or to delay the reward potential. I suppose that denying self-reward can, in some circumstances, actually be rewarding! Yet a further significant capacity of the human brain is that for introspection, not just in the context of resolving past unfortunate experiences, but also for aiding the success of future planning. Raising the introspective questions of 'why am I here?' and 'what happens when I die?' might explain why so many human societies have developed religious beliefs. It does not, however, explain why gods tend to be male or, indeed, why religious leaders and world leaders are also primarily male.

Past history would suggest that it was males who, without the burden of pregnancy or nurturing of babies, undertook the lead in acquiring land and possessions. Such activities further required leadership, a role based on command over social responsibilities, albeit of a more overtly assertive kind of command that sometimes engaged physical aggression and dominance. History has also shown that male leadership can produce a need to be controlling, and a resistance against reasoning. In the context of psychiatric disorders, such males are more likely to suffer bipolar disorders with grandiose and paranoid delusions, which can also result in an indifference to the suffering of others.

Pertinently, of the relatively few women who have featured in leadership roles, these few have rarely been shown to experience the extremes of mania that are seen to accompany male leadership. The striving for absolute power is not a feminine characteristic. And yet women can and do suffer from bipolar disorders, but interestingly these are principally of the depressive and anxious kind. The main psychiatric bipolar disorder of females is depression (Nivoli et al., 2011), while males have a greater incidence of manic episodes, paranoid personality disorders and social delusions (Diflorio & Jones, 2010).

Migration provides a further context for gender differences, principally influencing the well-being of women more than men, although social anthropologists have noted that the actual decision to emigrate was principally taken by males. Possibly because it was rarely their decision to emigrate, those women who did leave their natal social base had a higher risk of mood disorders. Women migrants show a lower self-esteem than male migrants and have a higher risk for depressive disorders (Giovanni et al., 2012). Moreover, neuroimaging studies of patients with depressive disorders show a sex-by-diagnosis interaction that influences the executive brain's structure (prefrontal cortex) and the functional brain areas that engage 'reward'. Activity in these brain regions are notably reduced in depressed female patients when they are given the task of assessing the emotions of other people from photographs, and when trying to determine if these images portray happy, angry or sad faces (Jogia et al., 2012). They are able to recognise faces, but not the kind of emotion they portray.

SIGNIFICANCE OF THE MATRILINE IN HUMAN SOCIETIES

Although it is matrilineal biology that has evolutionarily provided a strong predisposition towards social stability, anthropological studies suggest that matrilineal societies are extremely rare. African cultures have long believed that the mother and child relationship provides the foundations of society. Motherhood was a source of great celebration and core to the organisation of their communities. In Papua New Guinea, there remain both matrilineal and patrilineal societies whose main asset, land, is passed down through the female line and male line, respectively. A wedded woman in a patrilineal society is often from a neighbouring village and gains her knowledge of the husband's clan from her mother-in-law. Her status in the clan depends on her ability to raise sons and educate them about their land quality and distribution. In line with this, infanticide of daughters was often practised among these communities.

In the few matrilineal societies that have been available for study by modern-day anthropologists, women possessed land and were responsible for land decisions. Nevertheless, these decisions were often conveyed to the rest of their male society through a brother or an uncle. However, women in such matrilineal societies did not have to leave their clan, or learn a new way of life, or learn another clan's past history. Their understanding of, and their commitment to, social stability was reflected in the way they distributed the land, and their custom was to be fair to all. Matrilineal societies have, however, been relatively rare and patrilineality occurred five times more frequently than matrilineality among such island communities (Murdock, 1949). Many of the isolated matrilineal societies have, in recent decades, tended to revert to patrilineal, especially with respect to allocation of domestic resources, labour and capital. The purchase of brides (anthropologically referred to as 'bride price'), and hence male ownership of women, was often a feature of the social structure in patrilineal societies. Women carried out the drudge work, seed pounding and grinding, fetching water and firewood, and carrying infants and household possessions. Hunting with weapons was virtually a universal male speciality, and men monopolised weapons of war, which women were not even permitted to handle. Headmanship of the tribe (never a women) was widely accepted, and in the Shamanic tribes, leadership was always male.

A widespread cultural preference for male children has also been found to be predominant in patrilineal societies, and this was often embodied in a rule that the firstborn must be male. Demographic analyses taken before outsider contact was made, and when warfare was prevalent, shows that among those children under 14 years of age, boys outnumbered girls by 128 to 100, which could only have been achieved by female infanticide. The sex ratio of those children reaching 15 years or older was almost equal, around 50:50, primarily due to male mortality in warfare. Combat deaths could be blamed on the enemy, enhancing the social solidarity of the clan. Because of wars, female infanticide provided adaptive advantages for the clan demographic, primarily by maintaining sociosexual balance and stability among the population. The advantages to those clans which adopted warfare were that they did not suffer decreases in living standards, thereby avoiding hunger and disease, conditional on their being victorious. Societies which achieved stationary populations by other means were often defeated and destroyed by their aggressive neighbours.

A further example of sanctioned female infanticide has been found among the Netsilik Eskimos who live in the extremely harsh environments of north-west Canada (Balicki, 1970). Parents that were extremely loving and devoted to their children would sometimes leave their newborn daughter to die in the snow, notably if this daughter was born at a time when there was no male child of appropriate age (2–5 years old). A female without any prospects of marriage could not survive alone as an adult, and the alternative strategy of sharing very sparse resources with her could endanger the whole community. Thus, parents who were otherwise gentle and caring were nevertheless able to sacrifice their daughters for the long-term benefit of their community.

There are many examples of self-sacrifice for the benefit of others in society, none more widespread than those examples recorded in the context of warfare. Here the sacrifice is primarily made by males and again illustrates the regulatory power of mind (executive neocortex) over body. Interestingly, individuals have usually taken such selfless decisions under conditions of extreme physical and social hardship. Equally interesting, it is in such extreme contexts that the powerfully rewarding endogenous opioids, the body's 'painkiller' and the brain's 'feel-good factor' are released.

CONCLUDING COMMENTARY

Knowledge of our biological heritage is essential to understanding the predispositions and constraints which have selected for the evolutionary development of the human brain and, in turn, the influence this has had on culture and behaviour. Various mammalian species have gained restricted control over their environment by niche construction: otters building dams, rabbits digging burrows, bats nesting high in the trees and caves (Laland et al., 2016). One environmental niche which is common to all mammals is that of in-utero development. Not only does this provide a stable developmental environment in terms of oxygen, nutrient supply and constant temperature, but it embraces the coexistence of two genomes, maternal and foetal, thereby providing a potential platform for intergenerational co-adaptation. It is the mammalian brain which has been, and still is, the major beneficiary of the intergenerational co-adaptative process. This co-adaptive process takes place in the context of the early in-utero development by primarily involving the 'emotional' brain. Brain neurons are very energy-demanding and the larger the mammalian brain becomes, the greater are these energy requirements, and the longer it takes for the brain to develop. In-utero development therefore provides constraints on in-utero brain size, constraints which are of fundamental importance for limiting both the head size with respect to the birth canal and for delaying the huge energy demands which are required by cortical neural brain cells. Thus for a very large brain, especially that of humans, there is a second, even longer phase of brain development which occurs postnatally, primarily in an environment of matrilineal care, and subsequently in an environment that provides for social interactions within the community in which they live. These three niches – in-utero, maternal and social – are all important for the development of human well-being, but differ in their timing and consequences according to the stage of brain development. In-utero development requires a greater stability with a strong deterministic contribution from the genome, with emphasis on tight control over gene dosage such as that which is provided by genomic imprinting. The parts of the brain that are essential for breathing (oxygen), feeding (energy) and metabolism (temperature regulation) and the limbic brain's development of sex differences are the primary focus for early in-utero brain development. Extensive contributions from the pregnant mother, via the demands of the foetal placenta, ensure these early limbic areas of in-utero brain development receive sufficient energy resources. This further continues into the postnatal period, and again primarily via the milk which infants receive from maternal lactation. Interestingly, the continuation of maternal postnatal milk production is dependent on infant suckling, and may continue for years. This postnatal infant stage of survival depends not only on the essential contribution of the mother's milk, but on her warmth and protection against the climatic variance of the outside world. With increasing postnatal age, and particularly through the transitional years of puberty, the emotional secure attachment of infant with mother provides a secure platform from which the child's ever-expanding social world can develop. The important neural mechanisms for bonding, which evolved in the context of mother–infant development, are yet again to be deployed later in life for embracing the many other close relationships.

However, the developing male brain has an additional hurdle with which to contend, namely that of early masculinisation by testosterone. Studies reveal a negative social consequence of this masculinisation. This is exemplified by the male tendency to show less care-orientated and proactive behaviour in their social responsiveness than do females of the same age. It is therefore of no surprise that, in many primate societies, it is the males who leave their natal group and face a competitive struggle to enter another social group. Here males face the need to develop competitive behavioural strategies in order to remain and to achieve the social status required for their future mating potential.

The trajectory for primate brain evolution is thus determined by the need to develop and deploy intelligent behavioural strategies throughout developmental growth, and these evolutionary strategies differ for males and females. This brain growth progresses from the early in-utero stage, when brain development is primarily dependent on genetic determinants, to a brain that combines genetics and those epigenetic processes that are finely tuned by the social environment. The neocortex is a relatively slow-developing structure, and those parts which are uniquely enlarged in humans are the latest to develop, and continue to refine their interconnections in the adolescent period until 22 years of age. Thus, the maternal in-utero environment progressively becomes one small but essential part of an ever-expanding social world as brain development and male and female lifestyles progressively change.

An integral part of neocortical brain development is the inevitable appearance of developmental errors. These errors primarily result from those developing neurons which fail to make the right connections at the right time and consequently undergo programmed cell death. Cell death is a necessary and integral part of normal brain development. Of course, there is no hard-wired genetic programme which ordains when and where cell death occurs; this too is an epigenetic process initiated by a failure of developing neurons to reach their appropriate target at the right time. Moreover, a beneficial social environment is one that is essential to generating the kind of neural activity that is necessary for the guidance and consolidation of beneficial neural connection strengths. Conversely, the brain of infants reared in a socially deprived environment, such as that to which the Romanian orphans were subjected, fail to develop their optimal and appropriate neural connections. This type of social deprivation thus results in a brain that fails to consolidate those neural connection strengths which foster language development, and even the motor ability for infant advancement from crawling to walking.

Human postnatal brain development is especially notable for its engagement of the executive neocortex. This region of the brain continues to develop for the first 16 years of life after birth, followed by a further post-pubertal, adolescent period of synaptic pruning and neural interconnection refinement. The billions of neurons making trillions of synapses in the brain determine where the individual variance within the cortical brain structure is created. The consolidation of these synaptic connections and the future functioning of the executive brain in particular is therefore dependent on the environmental information which these regions of the brain encounter during this time period. Mother–infant bonding, and the security of infant attachment to the mother, thus provides the early foundations upon which later socially mediated epigenetic mechanisms shape the construction of an adult brain that is able to embrace and consolidate the complexity of future social life.

REFERENCES

Ainsworth, M. D. (1969). Object relations, dependency, and attachment: a theoretical review of the infant–mother relationship. *Child. Dev.* 40: 969–1025.

Aleksic, D., AsKsic, M., Radonjic, N. V., et al. (2016). Long-term effects of maternal deprivation on the volume, number and size of neurons in the amygdala and nucleus accumbens of rats. *Pyschiatr. Danub.* 28(3): 211–19.

Balicki, A. (1970). *The Netsilik Eskimo.* New York, NY: Natural History Press.

Bokhorst, C., Barkermans-Kranenburg, M., Fearon, P., et al. (2004). The importance of shared environment in mother–infant attachment: a behavior-genetic study. *Child Dev.* 74: 1769–82.

Bowlby, J. (1980). *Loss. Attachment and Loss Vol. II.* New York, NY: Basic.
Bowby, J. (1982). Attachment and loss: retrospect and prospect. *Am. J. Orthopsychiatry* 52: 664–78.
— (1991). Postscript. In: C. M. Parkes (ed.), *Attachment across the Life Cycle.* London: Routledge, pp. 293–97.
Broad, K. D., Curley, J. P. & Keverne, E. B. (2006). Mother–infant bonding and the evolution of mammalian social relationships. *Phil. Trans. R. Soc. B* 361: 2199–214.
Champagne, F. A. (2012). Interplay between social experiences and the genome: epigenetic consequences for behavior. *Adv. Genet.* 77: 33–57.
Chungani, H. T., Behen, M. E., Muzik, O., et al. (2001). Local brain functional activity following early deprivation: a study of postinstitutionalized Romanian orphans. *Neuroimage* 15: 1290–301.
Curley, J. P. & Keverne, E. B. (2005). Genes, brains and mammalian social bonds. *Trends Ecol. Evol.* 20: 561–67.
Diflorio, A. & Jones, I. (2010). Is sex important? Gender differences in bipolar disorder. *Int. Rev. Psychiatry* 22: 437–52.
Dixon, A., Dixon, B. & Anderson, M. (2005). Sexual selection and the evolution of visually conspicuous sexually dimorphic traits in male monkeys, apes, and human beings. *Annu. Rev. Sex Res.* 16: 1–19.
Fonagy, P. (1999). Psychoanalyctic theory from the viewpoint of attachment theory and research. In: J. Cassidy & P. R. Shaver (eds.), *Handbook of Attachment: Theory, Research, and Clinical Applications.* New York, NY: Guilford, pp. 595–624.
Giovanni, C. M., Francesca, M. M., Viviane, K., et al. (2012). Could hypomanic traits explain selective migration? Verifying the hypothesis by the surveys on sardinian migrants. *Clin. Pract. Epidemiol. Ment. Health* 8: 175–79.
Hesse, E. (1999). The adult attachment interview. In: J. Cassidy & P. R. Shaver (eds.), *Handbook of Attachment: Theory, Research, and Clinical Applications.* New York, NY: Guilford, pp. 395–433.
Insel, T. R. & Young, L. J. (2000). Neuropeptides and the evolution of social behavior. *Curr. Opin. Neuorbiol.* 10: 784–89.
Jogia, J., Dima, D. & Frangou, S. (2012). Sex differences in bipolar disorder: a review of neuroimaging findings and new evidence. *Bipolar Disord.* 14: 461–71.
Keverne, E. B. (2002). Pheromones, vomeronasal function, and gender-specific behavior. *Cell* 108: 735–38.
— (2005). Neurobiological and molecular approaches to attachment and bonding. In: C. S. Carter et al. (eds.), *Attachment and Bonding: A New Synthesis.* Cambridge, MA: MIT Press.

Kim, S., Kwok, S., Mayes, L. C., *et al.* (2016). Early adverse experience and substance addiction: dopamine, oxytocin, and glucocorticoid pathways. *Ann. N.Y. Acad. Sci.*

Kraemer, G. W., Ebert, M. H., Schmidt, D. E., *et al.* (1991). Strangers in a strange land: a psychobiological study of infant monkeys before and after separation from real or inanimate mothers. *Child Dev.* 62: 548–66.

Laland, K., Mathews, B. & Feldman, M. W. (2016). An introduction to niche construction theory. *Evol. Ecol.* 30: 191–202.

Lyons, D. M. & Schatzberg, A. F. (2003). Early maternal availability and prefrontal correlates of reward-related memory. *Neurobiol. Learn. Mem.* 80: 97–104.

Martel, F. L., Nevison, C. M., Rayment, F. D., *et al.* (1993). Opioid receptor blockade reduces maternal affect and social grooming in rhesus monkeys. *Psychoneuroendocrinology* 18: 307–21.

Martel, F. L., Nevison, C. M., Simpson, M. D. A., *et al.* (1995). Effects of opioid receptor blockade on the social behaviour of rhesus monkeys living in large family groups. *Psychobiology* 28: 71–84.

Matsumoto, K. & Tanaka, K. (2004). The role of the medial prefrontal cortex in achieving goals. *Curr. Opin. Neurobiol.* 14: 178–85.

Mori, K., Fujita, S. C., Imamura, K., *et al.* (1985). Immunohistochemical study of subclasses of olfactory nerve fibers and their projections to the olfactory bulb in the rabbit. *J. Comp. Neurol.* 242: 214–19.

Murdock, G. P. (1949). *Social Structure*. New York, NY: Macmillan-Collier.

Nivoli, A. M., Pacchiarotti, I., Posa, A. R., *et al.* (2011). Gender differences in a cohort study of 604 bipolar patients: the role of predominant polarity. *J. Affect. Disord.* 133: 443–49.

O'Connor, T. & Croft, C. (2001). A twin study of attachment in preschool children. *Child Dev.* 72: 1501–11.

Porreca, A., De Palo, F., Simonelli, A., *et al.* (2016). Attachment representation and early interactions in drug addicted mothers: a case study of four women with distinct adult attachment interview classifications. *Front. Psychol.* 7: 346.

Ranote, S., Elliott, R., Abel, K. M., *et al.* (2004). The neural basis of maternal responsiveness to infants: an fMRI study. *Neuroreport* 15: 1825–29.

Santona, A., Tagini, A., Sarracino, D., *et al.* (2015). Maternal depression and attachment: the evolution of mother–child interactions during feeding practice. *Front. Psychol.* 6: 1235.

Stowers, L., Holy, T. E., Meister, M., *et al.* (2002). Loss of sex discrimination and male-male aggression in mice deficient for *TRP2*. *Science* 295: 1493–500.

Striedter, G. F. (2005). *Principles of Brain Evolution*. Sunderland, MA: Sinauer Associates, Inc.

van IJzendoorn, M. H. & De Wolff, M. (1995). Adult attachment representations, parental responsiveness and infant attachment: a meta-analysis on the predictive validity of the Adult Attachment Interview. *Physiol. Bull.* 117: 387–403.

Vines, G. (1996). Between heaven and hell. *New Scientist* Supplement April: 22–23.

Young, L. J. (1999). Frank A. Beach Award. Oxytocin and vasopressin receptors and species-typical social behaviors. *Horm. Behav.* 36: 212–21.

6 Brain and Placenta: The Coming Together of Two Distinct Generations

Two major biological developments have taken place in the evolution of mammals, each of which has produced a significant impact on the genetic regulation of brain development and behaviour. The first of these major mammalian developments was giving birth to live babies (viviparity) as opposed to egg-laying and their external incubation, a common form of reproduction for the great majority of our vertebrate ancestors. Pivotal to, and essential for, the success of mammalian viviparity has been a uniquely mammalian form of development, namely that of the foetal placenta. Evolution of the placenta enabled mammalian foetal development to occur within mother which placed a considerable burden of both time and energy demands upon mother. Moreover, this maternal commitment continues into the postnatal period, thus ensuring a sustained investment of the mother's time and energy being dedicated to foetal development following the infant's loss of the placenta at birth. These events have resulted in essential modifications to the mother's 'limbic emotional' brain, not only in the context of her decreased motivation for sexual behaviour, but also for the onset of parturition, milk production, maternal feeding, maternal care and the suspension of maternal fertility. Such activities have been evolutionarily integral to the success of viviparity in the majority of mammalian species. Both pregnancy and postpartum care have thus required adaptive evolutionary changes to the maternal brain. Indeed, control over these reproductive events is primarily a function of the mother's limbic 'emotional' brain (notably the hypothalamus and its interconnections), which has evolved to respond to foetal needs via those hormonal instructions produced by the placenta of the foetus (Keverne, 2006). Placental viviparity has

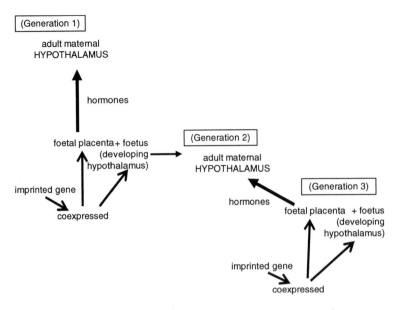

FIGURE 6.1 Co-adaptive evolution over three generations of genomic imprinting. Imprinted genes are coexpressed in the foetal placenta and foetal hypothalamus (Table 6.1) during the developmental period when the foetal placenta instructs the mother's hypothalamus for maternal behaviour. The foetal–placental transgenerational engagement of the mother's hypothalamus is essential for the successful destiny of its own generation. Thus, a transgenerational template is provided for selection pressures to evolve successful development for foetal placental–maternal hypothalamic transgenerational interactions.

thus introduced an important new dimension with respect to mammalian reproduction and involving the coexistence of matrilineal genomes over three generations (mother, foetus, and in the case of daughters, the developing oocytes within the foetus that will produce the next generation). Hence, the successful reproductive life of the female, but emphatically not that of the male, engages three generations, all of which are integral to the mother and her foetus (Figure 6.1).

Successfully giving birth to live babies (viviparity) is no simple matter, and its success has primarily depended on the mother's integrated response to the development of the placenta. The

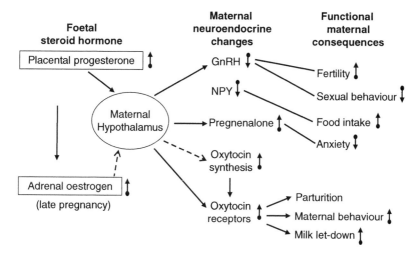

FIGURE 6.2 The effect of foetal steroid hormone progesterone. The placentally produced steroid hormone progesterone determines forward maternal planning by directing/orchestrating maternal physiology and postnatal maternalism to synchronise with the development of the foetus.

foetal placenta is genetically distinct from the mother, and serves to transfer maternal nutrients and blood-born oxygen to the foetus, as well as removing waste products. Even more evolutionarily thought-provoking is the way this next generation's foetal placenta has produced hormonal messengers that successfully instruct the mother's brain to 'switch off' her reproductive system, including her sexual activity. It is also these placental hormones which actively prime the mother's brain to be maternal in readiness for the birth of the baby. As part of this priming process, the placental hormones stimulate increased maternal feeding to boost the energy required, especially for the later stages of pregnancy and for maternal milk production (Figure 6.2). Because the placenta is foetal in origin, then many of its cell types run the risk of being rejected by the mother's immune system, in a way similar to the rejection mechanism which occurs with non-self organ transplants. Successful pregnancy has thus required complex localised co-adaptations between the foetal placenta and the mother's immune system in order to avoid foetal rejection (see Conclusion for more detail). Placental viviparity has further introduced an important reproductive advantage that has resulted from the coexistence, over three generations, of matrilineal genomes.

In parallel with placentation has been the evolution of maternal epigenetic marks located to DNA which serve as gene expression control regions for genomic imprinting. These maternal heritable epigenetic marks undergo reprogramming in the developing maternal oocyte of each new generation, thereby ensuring continuity of imprinted gene expression according to the maternal parent of origin for each successive generation (see Chapter 3). Hence, the action and interaction of epigenetically imprinted gene expression has outcomes that involve expression of this gene's same allele across two genomes (maternal and foetal). It is this intergenerational continuity of expression for imprinted genes which provides a template for intergenerational selection pressures to operate, particularly via the matriline. This maternal hypothalamic–foetal placental template has been essential for shaping the mothering capabilities of each subsequent generation. Such intergenerational genetic co-adaptations between the maternal hypothalamus and foetal placenta have thereby provided the integrative platform on which selection pressures can and have operated for the successful evolution of mammalian reproductive viviparity (Figure 6.3).

In conclusion, the matrilineal contribution to mammalian reproductive success has depended on the intergenerational co-adaptive evolution of the brain and placenta. This co-adaptation underpins a common evolutionary goal of coordinating mother–infant physiological homeostasis, ensuring maternal energetic investment, and providing focussed maternal care. The evolutionary outcome of this linked co-adaptation is that infants that have extracted optimal maternal nurturing will be well provided for, and both genetically and epigenetically predisposed towards good mothering in the next generation.

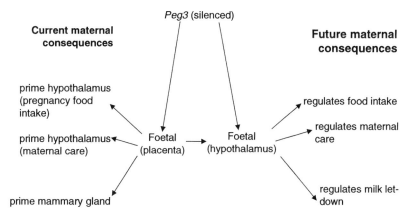

FIGURE 6.3 Co-adaptive functioning of the *Peg3* imprinted gene. The similarity of functional phenotypes with *Peg3* mutation in the developing placenta or independently in the developing hypothalamus illustrates its co-adaptive functions across generations.

BRAIN EVOLUTION

Within and across mammalian species, a significant and progressive evolutionary enlargement of the 'executive' brain has taken place. This large brain has substantially emancipated much of primate, and especially human, adult-motivated behaviour from the deterministic effects of hormones that are essential for maternal care in small-brained mammals. It is this growth of the executive brain that has played the latest and most significant role in shaping the evolutionary trajectory of humans. The development of such a large, adult executive brain has occurred under selection pressures for coping with the social environment and this has, in turn, resulted in the emancipation of brain development from those physiological hormonal messengers that determine behaviour in small-brained mammals. The human executive brain depends upon the deployment of both genetic and epigenetic strategies in its development. This is brought about primarily through the engagement of the maternally imprinted genes and their transgenerational expression, employing the same genetic allele in determining both the development of the foetal placenta and the foetal brain (Keverne, 2013).

Emancipation of the executive brain from the hormonal determinants in Old World primates and humans has thus enabled the mother's younger sisters and daughters to participate in infant care before they reach reproductive age. In this way they enhance their parenting skills under the watchful eye of the mother before they themselves undertake their hormonal priming for the next generation's pregnancy. In essence, the evolution of the executive brain has shaped the trajectory for its own destiny. However, we must not underestimate the significant part played by the matriline in achieving this evolutionary trajectory. In short, we can place these executive brain advances as being consequential on successful placental viviparity. Placental viviparity has provided the template for selection pressures to successfully operate in determining intergenerational co-adaptation between the brain and the placenta (Figure 6.1).

The brains of different mammals, when taking into account their owners' body size, show a wide variation in size that accords with their lifestyles. Primates (monkey, apes and humans) stand out as having exceptionally large brains relative to body size. Among these primates, humans are particularly exceptional in possessing a brain three times the size of our nearest primate relative, namely the chimpanzee (Striedter, 2005). In absolute terms the elephant's brain is larger than ours, but so too is the elephant's body, which from an evolutionary perspective is why it is necessary to take into account body weight when comparing the relative size of brains across different species. There are a number of approaches to exploring mammalian brain evolution including consideration for the complexity of lifestyle, which evolutionary biologists have suggested favours the evolution of larger brains. It has also been hypothesised that selection pressures for certain cognitive skills have arisen from ecological variables such as diet and home range size. It is certainly the case that knowledge of available nutritional resources and the optimal time for their eating, together with knowledge of their spatial distribution, requires cognitive brain maps. However, this is but one of the many variables to be considered in the context of brain evolution, and other studies have placed more emphasis on the complex nature of social life, which is also an important feature of many primate societies.

While the size of the brain as a whole has been linked to these different lifestyles and ecological features, this approach fails to take into account the heterogeneous structure of the brain and the different functions served by its component parts. Thus, the size of the executive neocortex relative to overall brain size is significantly larger in social primates compared with solitary species, while among primates in general the size of the neocortex, taking into account body weight, varies with social group size (Striedter, 2005, ch. 8). Larger social groups present greater social complexity, which in turn requires an executive brain capable of understanding and sustaining the greater social knowledge such a lifestyle embraces.

Examining the skull and cranial cavity from our human ancestors gives some gross estimates of their brain size, although it is by no means clear that the human brain has evolved as a gross unitary structure. For this reason, it is more appropriate to consider the functional divisions of the brain, as some brain regions are notably larger than others. In terms of function, the most anterior part of the human executive brain, the frontal neocortex, functionally provides insight and a capacity to identify with other individuals, a process which psychologists have named 'theory of mind'. The frontal neocortex is also essential for successful functioning in the context of anticipation and forward planning. The frontal cortex does not, however, function in isolation and is intimately related to other areas of the cortex and, in particular, to the striatum. The striatum is a deep structure of the brain surrounded by the neocortex and plays an important role in transforming cortical instructions into actions, and to providing the reinforcing 'neural reward' for those actions that prove to be advantageous (Schultz, 2000). These two brain regions (neocortex and ventral striatum) represent the areas of the brain which have expanded most during human evolution, and are collectively referred to as the 'executive' brain.

From a behavioural viewpoint, other parts of the brain might be considered to potentially conflict with the decision-making process of the 'executive' brain. Certain areas of the limbic brain (hypothalamus, septum and preoptic area) are important for primary motivated behaviour such as sexual, aggressive, feeding and parental behaviour. For the purpose of this account, I have grouped these brain regions as 'emotional' to distinguish them from the executive neocortex. These areas of the 'emotional' brain are under strong hormonal influences, and the decision-making process for reproduction and reproductive behaviour is principally determined by this neurohormonal relationship in those mammals with a relatively small executive brain.

Underpinning successful brain evolution has required the collective activity of many genes as well as the collective involvement of functionally distinct areas of the brain, both executive and emotional. These distinct brain regions also form natural groupings in relationship to what we know about the role of key regulatory genes in mammalian brain development. Developmental disorders of the human brain would also suggest these 'executive' and 'emotional' descriptors for the brain are genetically valid from what we also know about the genetics underpinning human brain disorders. An appropriate example of such genetic brain dysfunction is the Prader–Willi syndrome and Angelman's syndrome. Children with Prader–Willi syndrome are voracious eaters, develop an obese phenotype and are prone to temper tantrums followed by remorse (emotional brain). Children with Angelman's syndrome have severe intellectual disability, speech impairment, epilepsy and movement disorders (executive brain). Interestingly, the genetics of both disorders are underpinned by genes which have undertaken epigenetic regulation through genomic imprinting, with impairment for maternal expression of alleles being responsible for Angelman's syndrome (executive brain) and impairments to paternal allele expression being responsible for Prader–Willi syndrome (emotional brain).

CO-ADAPTIVE DEVELOPMENT OF THE MATERNAL 'EMOTIONAL BRAIN' AND THE FOETAL PLACENTA

Mammalian viviparity requires a foetal placenta that physically invades the mother's uterus, thereby resulting in the coexistence of two generations with different genomes, maternal and foetal, each interacting and requiring co-adaptation across these two generations (Keverne, 2015). The success of this co-adaptive development of mother's hypothalamus and the foetal placenta is dependent on the intergenerational signalling by the foetal placental hormones to the neurons in the mother's emotional brain (hypothalamus). In this way the genome of the foetal placenta, by way of controlling the production of placental hormones, directs multiple aspects of the mother's 'emotional' (hypothalamic) brain function (Figure 6.2). Placental hormones increase maternal feeding, one of the earliest signs of pregnancy, and this takes place in advance of the energy requirements needed for that greater part of foetal growth which occurs later in pregnancy. Were it not for this advanced accumulation of energy reserves, the pregnant female would fail to cope with those later increased nutritional requirements for the subsequent increased growth demands of the foetus. The hormones of the foetal placenta also terminate maternal fertility and suppress sexual behaviour in the majority of mammalian species (humans are an exception), thereby ensuring that maternal time and energy expenditure is directed towards the success of pregnancy. Other placental hormones induce maternal nest-building in advance of birth, and also prime the mother's hypothalamus for postpartum maternal care and the mammary glands for sufficient milk availability (Figure 6.2).

In short, the genome of the foetal placenta controls the destiny of the foetus by producing hormones that interface with mother and instruct successful maternalism via her limbic/emotional brain. Because the in-utero foetus is developing its own emotional brain during this period of co-adaptation with the mother, an intergenerational template is made available on which selection pressures may operate intergenerationally across their genomes (Figure 6.1). So efficient has this intergenerational co-adaptation been that female infants, who have obtained optimal nourishment and care from their mother, will themselves be both genetically and physiologically predisposed to good mothering for the following generation. Most infants do receive devoted care, and are thereby further advantaged for an early puberty as a booster to their lifetime's reproductive success. Conversely, severe maternal stress and the maternal hormonal changes induced by stress may also influence changes in placental function which adversely influence foetal 'programming' for the next generation. Such adverse maternal stress-induced foetal programming is well recognised in the context of generating future hypertension and metabolic disorders. Interestingly, the foetal placenta can also produce hormones that are anxiolytic and which ameliorate any mild stress, thereby having a calming effect on the mother.

How these advantageous intergenerational co-adaptive processes evolved is an interesting question, and one which has also been addressed by molecular genetic studies. Such studies have identified a number of imprinted genes and their downstream networks which are expressed in the developing emotional brain (hypothalamus) of the foetus, and are also coexpressed, at the same time, in the developing placenta of this foetus (Table 6.1). Deleting the expression of such imprinted genes in either the placenta of the current generation or in the developing foetal hypothalamus of the next generation produces remarkably similar functional impairments (Figure 6.3). The success of this next generation's growth and maternal care is impaired by maternal hypothalamic gene deletion, while provisioning for offspring growth via the placenta and maternal behaviour of the present generation is disrupted via a failure of placental hormone production (Keverne, 2015). Let me explain this in a little more detail. First, consider a mutation to an imprinted gene that is selectively expressed in the mother's brain (hypothalamus) of the current generation, but is not expressed in the foetal placenta (next generation). This matrilineal hypothalamic genetic impairment reveals that many of the phenotypic outcomes have similar functional consequences to the same mutation of this same imprinted gene, but selectively restricted to the foetal placenta (next

Table 6.1 *The foetal placenta exhibits forward planning by regulating the maternal hypothalamus.*

(a) by increasing maternal food intake before there is energetic demand
(b) by suppressing sexual behaviour and fertility
(c) by initiating nest-building before offspring birth
- **Question**

How has the maternal hypothalamus evolved to plan ahead for these adaptations that respond to the foetal placenta of the next generation?

---

generation). These studies thus demonstrate that a placental gene mutation is indirectly influencing the functioning of the mother's brain (hypothalamus) via the impairment of placental hormone production (Curley & Keverne, 2005) (Figure 6.3). Moreover, the functional consequences of the gene deletion made directly to the mother's hypothalamic brain is manifest by a reduction in the number of neurons that survive here during development. This, in turn, results in impaired hypothalamic ability for the mother to respond to the placental hormones that determine her maternal care for the next generation, and further produces neural impairments for milk let-down. Moreover, deletion of this same gene selectively in the foetal placenta but not in the maternal brain impairs both placental growth and the production of those placental hormones which are required for priming of the mother's hypothalamic brain which determines maternal care and milk let-down. The consequent low birthweight of her infant further impairs postnatal infant growth and delays puberty onset for this following generation. Such functionally convergent actions of imprinted genes across the two generations (emotional brain in the mother, placenta in the foetus) illustrate exactly how successful these co-adapted imprinted genes have become. They have evolutionarily determined the intergenerational co-adaptation between the mother's emotional brain and the foetal placenta, which has been achieved via the placental hormonal messengers of the foetal placenta controlling the mother's behaviour and physiology.

Interestingly, during this critical in-utero period for development of the foetal emotional brain, some of those imprinted genes which change their expression in the developing placenta also change their expression in the developing emotional brain (Table 3.1). In this way, transcriptional synchrony is ensured across the brain and placental co-adapted tissues. Moreover, during this same developmental time period, there is an increase in the number of downstream non-imprinted genes which also show coordinated changes in their placental expression with changes to the foetal emotional brain coexpression (Keverne, 2015). These are the same genes that are most affected by mutations across the imprinted gene networks (Varrault *et al.*, 2006). Hence we are presented with a clear demonstration for the significance of synchronised patterns of gene transcription which produce changes for developing the future mother's brain and for the current foetal placenta. Desynchronisation of this multiple gene coexpression, which occurs when the co-adapted imprinted gene is mutated, produces functional impairment to the developing maternal brain for the next generation, and impairment of placental functioning in the present generation. These studies illustrate the importance for such co-adapted intergenerational evolution provided by epigenetically imprinted genes. Moreover, it is this early developmental time period which is clearly important for co-adaptive selection pressures to operate during the development of the foetal emotional brain (hypothalamus) and foetal placenta. In this way, the matching of the mother's expressed allele with the offspring's expressed allele at those imprinted loci that effect both maternal and offspring traits benefits their interaction. This is not only the case for the many imprinted genes that are coexpressed in the mother's hypothalamus and foetal placenta, but also for other imprinted genes that are coexpressed in the mother's mammary glands, and also affect the infant brain's control of suckling in the offspring (Cowley *et al.*, 2014).

The success of mammals in producing thriving babies has thus depended on synchronisation of genetic events across two generations, involving the mother's emotional brain (hypothalamus) and the foetal placenta. Errors in gene expression occurring in either generation can spell disaster. However, because the placenta and hypothalamus are co-adapted, the placenta has the potential to make short-term sacrifices should problems arise that might compromise development for the next generation's developing brain (hypothalamus). Hence, a period of food deprivation, late in pregnancy, has potential problems for the development of neural connections that occur at this time within the foetal hypothalamus. Interestingly, experiments have shown that in order to overcome any lack of nutrients, the placenta itself undertakes autophagy, a kind of limited self-digestion, to provide the nourishment which the developing foetal hypothalamus requires (Broad & Keverne, 2011). The placenta, at this late developmental stage, will, in any case, terminate within a few days, but for the developing hypothalamus, this still requires further energy for the neural interconnections to be made with the rest of the brain. Placental autophagy thus provides the essential proteins to avoid any negative outcomes that may arise for the next generation, and that result from a shortage of maternal nutrition late in pregnancy. This may also explain why most mammalian mothers eat the placenta following birth to ensure maternal vital energy requirements are restored following parturition.

These studies clearly illustrate exactly how important intergenerational co-adaptation has been for mother and foetus. By presenting the two genomic generations as a single genetic unit during in-utero foetal development, co-adaptation has provided a new dimension on which natural selection may operate. Brain and placental co-adaptation has thereby ensured that rapid and positive intergenerational evolutionary progression can be facilitated. These studies also raise the question as to why sons, who also possess a developing emotional brain and placenta, fail to become maternal when adult. Why is it that the placental hormones fail to shape the maternalisation

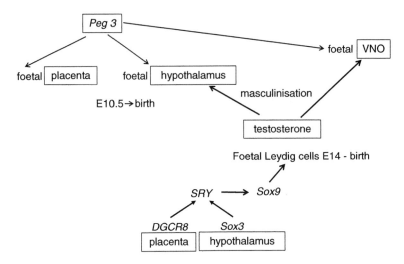

FIGURE 6.4 Patriline masculinisation. The imprinted gene (*Peg3*) is important for the co-adaptive development of the foetal placenta and foetal hypothalamus, together with the foetal vomeronasal organ (VNO). During this period, the foetal Leydig cells produce testosterone which masculinises the male hypothalamus and vomeronasal pheromone pathway. The foetal Leydig cells are produced via *SRY* activating *Sox9*. *SRY* is the male sex-determining gene produced by fusion of *Dgcr8* (also expressed in the placenta) and *Sox3* (expressed for developing hypothalamus).

of the male hypothalamus? To answer this question it is first necessary to consider the developmental process of masculinisation for the male's emotional brain (hypothalamus) and the production of male hormones that are required for this process.

Males have a Y-chromosome that provides for gene expression that ensures testicular development (see Chapter 1). The testes produce the masculinising hormone testosterone which acts on the developing brain to ensure males are both sexually active and aggressive, but prevents them from becoming maternal. Masculinisation has thus uncoupled the synchronised co-adaptive development of the hypothalamus and placenta in the foetal male (Figure 6.4) while not uncoupling the foetal male's own placental hormones from interacting with the mother's mature hypothalamus to induce her maternal care of male infants as well as female infants. In this way males, like the females, have benefitted from maternal care via their placenta, but unlike females, males of most mammalian species undertake testosterone masculinisation of their hypothalamus which has prevented them becoming maternal in the next generation. In summary, this biological barrier to the provisioning of infant care by males operates at two distinct levels. First, by the masculinisation of the male's developing foetal brain, and second by the absence of a placenta and thereby the hormones it produces. Hence the production of those placental hormones which act on the female's (mother's) hypothalamus are not available for the adult male hypothalamus. Unlike the adult female, the adult male is never pregnant and is therefore never exposed to placental hormones which induce maternal behaviour. The adult male hypothalamus thus lacks any possibility for intergenerational co-adaptation with the foetal placenta. Moreover, adult males never experience parturition (the birth process) which provides the activator signal to the mother's brain for induction of oxytocin release, and hence the immediate promotion of maternal behaviour at birth. Indeed, the males of most mammalian species are rarely present during the birth process, with the exception of a very few monogamous species.

A further development that is special and different for the male is the precision timing for masculinisation of the male brain. Thus, in parallel with the evolution of female in-utero development, has been the evolutionary origins of the mammalian male testes-determining *SRY* gene.

MASCULINISATION OF THE MALE HYPOTHALAMIC BRAIN

The *SRY* gene is the master switch for male sex determination, but this gene is not present in the earliest non-placental mammals. Sex determination for non-placental mammals requires the activity of multiple genes across many chromosomes. Moreover, the mammalian *SRY* is not a new gene, but a hybrid of two genes, *Sox3* and *Dgcr8* (*Di-George Critical Region 8*). *Sox3* is required for, and is integral to, formation of the hypothalamic emotional brain and pituitary gland in both males and females (Sato et al., 2010). The pituitary gland is the emotional brain's interface for regulating the male and female gonads, while the emotional brain's hypothalamic neurons are integral to both male sexual behaviour and female maternal care. *Dgcr8*, the gene which has combined with *Sox3* to produce the *SRY* gene, is itself integral to developing the placenta (Figure 6.4). Foetal males also have a placenta with the capacity to produce the hormones that activate their mother's maternal care. The *SRY* gene's ability to initiate the development of male foetal Leydig cells, and in turn their foetal production of testosterone, prevent any future 'maternal' responsiveness by males as a consequence of masculinisation of their developing brain (hypothalamus) (McCarthy, 2010). The developmental timing for this masculinisation of the male's brain is critical to prevent the activation of genes which would otherwise establish a maternal brain. The male genome thus appears to have achieved masculinisation not by the creation of any new genes, but by hybridisation of already established genes to produce the new *SRY* gene. *SRY* activates and synchronises downstream gene transcription to ensure development of the foetal Leydig cells. These cells produce their foetal testosterone in the absence of any intermediary pituitary hormones, and at the critical time for determining developmental masculinisation of the hypothalamus, which takes place during male in-utero development (see Chapter 1).

Brain development and brain evolution have indeed been a complex process which has taken place in stages over many millions of years. The non-placental mammals are thought to have split from the mammal-like reptiles some 200 million years ago. The first placental mammals, the marsupials, diverged from the egg-laying monotreme mammals about 130 million years ago, but the origin of the invasive placental mammals diverged only some 20 million years ago. Among these invasive placental mammals, it is the executive brain which has increased in size across different phylogenies, but most notably in the primates (monkeys, apes and humans). The emotional hypothalamic brain has diminished in relative size over this same period for these same species, thereby providing clear evidence for the mammalian brain not having evolved as a unitary structure. Furthermore, other evolutionary developments have coincided with the increases in executive brain size, namely an emancipation from the deterministic hormonal regulatory control over behaviour.

In the evolutionary context of the primate executive forebrain remodelling, those areas of relative increased brain growth are those to which the maternal genome (as determined from matrilincal parthenogenetic distribution of chimeric cells) makes a substantial developmental contribution. This matrilineal genomic contribution is especially noteworthy with respect to understanding the evolutionary increase in size of the primate forebrain, and the functional role this has played in the development of complex social strategies. The way in which such social strategies are deployed is especially notable for the matriline, and complex primate societies are often described as being 'female-bonded' (see chapter on bonding). Regions of the brain which reveal relative evolutionary decreases in size are those to which the paternal genome, through genomic imprinting, makes its differentially greater contribution (Figure 3.6, Chapter 3). It is these emotional brain regions that are the targets for both the gonadal and the stress hormones. Such evolutionary progression for relative size reduction of the emotional brain is congruent with the emancipation of primate reproductive behaviour from the deterministic hormonal regulatory control in the adult. These findings are also important for understanding the significance of forebrain expansion in the development of complex social behaviour. This is especially the case for human societies, where social structure takes the lead as the major determinant for many forms of behaviour. In other primate species, maintenance of social cohesion and group continuity over successive generations is dependent on the matriline, and it is the females of high social rank that produce high-ranking daughters that stay within their matrilineal group (Broad et al., 2006). Brain areas that have decreased in relative size over the same evolutionary timescale are those areas to which the father's genome makes a differentially larger contribution, areas which are the target for gonadal and stress hormones. Overall, these findings are congruent with a diminished role for gonadal hormones being the determining factor for reproductive behaviour, especially as progression for a larger neocortex and complex social organisations evolved. Reducing the significant role for signalling via hormones in determining reproductive success within primate societies has been succeeded by the importance of social status. An integral component of primate social status is the process of learning and memory, and with it the deployment of intelligent behavioural strategies under control of the executive brain.

THE MATRILINEAL IMPORTANCE FOR NEOCORTICAL BRAIN EVOLUTION

Mammals have shown an evolutionary logarithmic increase in brain size relative to other vertebrates, and this increase is even greater within and across mammalian phylogenies than that which occurs between early mammals and their ancient reptilian ancestors. Such exponential cortical growth makes the mammalian brain very special, and its size increase, compared with body weight increase, reaches a peak in the primates and especially humans. Primates with a large executive brain differ from other mammals in the complexity of their social interactions, while their primary motivated behaviour (sex, aggression, maternal behaviour) is less dependent on the determining effect of hormones acting on their emotional brain. Thus in large-brained primates parenting is not restricted to the post-partum period, and sexual interactions are not restricted to a discrete period of oestrous in these species. This is primarily a consequence of their large neocortical expansion and the greater role for this part of the brain in reproductive decision making. The fact that the primate's neocortex continues to develop long into the postnatal period, coinciding with the timing of close relationships with the mother and with other members of their social group taking place, has been hypothesised to be the most important socioselection factor in primate brain evolution (Keverne et al., 1996).

Among these highly social Old World monkeys, it is the females rather than the males who determine the group's social stability and cohesion. This is maintained over successive generations, with the social rank of daughters being inherited from their mother. Again, it is the matriline that has been pivotal in determining such social evolution, and the neural mechanisms subserving this social evolution engage those regions of the brain that are also involved with mother–infant bonding. The functioning of these same neural circuits is further linked to the biology of brain reward. Moreover, the affiliative behaviour which underpins social reward, such as grooming and sexual behaviour, results in the same neuropeptide hormone transmitters (oxytocin and β-endorphin) being released in the brain when the mother gives birth (Curley & Keverne, 2005). These neuropeptides have acquired the distinctive function of rewarding positive social encounters, thereby providing the 'social glue' for complex primate societies as well as for mother–infant bonding. In order for social reward to have become such an important selection factor in neocortical brain evolution, it was initially important that maternal care became emancipated from a deterministic action of hormones on the brain. Thus in primate societies, it was this emancipation from hormonal regulatory determinants that provided the possibility for the older sisters and maternal relatives to also participate in offspring care without themselves having undertaken pregnancy or parturition. Such social alloparenting also provided helpful assistance to the mother, thereby liberating her from delaying her next pregnancy for the many years it takes until her offspring's brain is fully mature. The extended matrilineal family fulfils this care-giving role, and surveyed by the watchful eye of the matriarch, older daughters gain experience in mothering before they themselves undertake pregnancy. With the evolutionary enlargement of the executive primate brain, social learning has epigenetically replaced the deterministic role of hormones in assuring the success of maternal care. Of course, hormonal changes still remain integral to parturition and to lactation, but are no longer such an integral requirement for primate, especially human, maternal care. This emancipation of human behaviour from hormonal determinants has enabled women to adopt children and become perfectly good mothers. Unfortunately, such emancipation also means that some mothers may neglect their baby, even though they have undertaken a successful pregnancy without any hormonal irregularities.

Epigenetic processes (see Chapter 2) in the brain are known to be activated by prosocial mothering (high levels of licking and grooming) in rodents, and produce offspring that display similar prosocial mothering in the next generation, even when these offspring are not genetically related to this surrogate mother. Likewise, studies involving mother–infant separation and maternal deprivation in monkeys that were carried out by Harlow in the early 1950s in the USA found that maternal deprivation had severe consequences for mothering by daughters of their next generation. These studies illustrated the importance of a mother's interaction with her infant for the future normal behavioural development of these infants. Infant rhesus monkeys reared in the absence of mother are behaviourally inhibited, show increased sensitivity to stressful events, and display impairments in social and reproductive behaviour when adult (Hinde et al., 1978). Likewise, early social deprivation of institutionalised human infants results in impulsivity, cognitive impairment, attentional and social deficits in later life. This is characterised by dysfunctional neural activity in the executive brain (orbital frontal cortex, prefrontal cortex and medial temporal cortex) and is underpinned by failures to make the appropriate interconnections of the executive brain, as revealed by neural tract imaging studies (Eluvathingal et al., 2006). How these early influences on brain development produce such long-term changes in the behaviour of children is not well established. However, because brain development is activity-dependent, the epigenetic consequences of social neglect may impair the appropriate neural connections and activate inappropriate, maladaptive neural interconnections (see Chapter 2 on epigenetics).

Studying the genetic basis of mammalian brain evolution and the genes which underpin brain differences across primate species has met with little success. The human executive brain is three- to fourfold the size of that found for the chimpanzee brain, but human/chimpanzee brain-expressed gene comparisons show remarkably strong genetic sequence alignments. This makes any estimates for the evolutionary rates of change for these nervous system genes difficult to assess. Although the brain gene sequence differences between human and the chimpanzee are marginal, some 44 per cent of brain genes nevertheless show evidence for some form of differential timing for their expression across the brains of these species. Recent molecular genetic studies have shown that it is the non-protein-coding regions of the human genome which display the most notable evolutionary acceleration of substitutions. Moreover, these non-coding regions of the human genome do indeed diverge from those of the chimpanzee. Many of these genes are non-protein-coding RNA elements that serve as epigenetic regulators of gene function, important for determining gene transcription during neocortical development. Although brain protein-coding genes tend to be highly conserved, there is a subset of the non-protein-coding regulatory RNA sequences which have been found to have undergone human-specific accelerated evolutionary change (Pollard *et al.*, 2006).

The complexity of the human executive neocortex raises a further important question as to how billions of neurons and trillions of synapses are successfully brought together as functional units, using this fairly restricted number of available genes. Moreover, as already mentioned, the number of protein-coding brain genes involved in human executive brain development appear to be only marginally different from those of the chimpanzee, which has a neocortex about one-third the size of the human neocortex. In addition, the human brain has not only become evolutionarily larger than that of the chimpanzee, but it has also become more internally reorganised. Thus the ratio of executive brain to brain stem is twice as large in humans as it is in the chimpanzee, and as the brain stem itself is unlikely to have decreased as humans evolved (indeed, it too has increased relative to the chimpanzee's), then the human neocortex has enlarged disproportionately. Investigating this kind of basic comparative neuroanatomy leads to the conclusion that it is the regulatory 'time and space' changes to gene expression rather than any increases in the number of different protein-coding genes that have played the most significant role in human brain development and evolution. This finding is encapsulated not in single genes, but in the increased complexity of overlapping gene networks and their epigenetic regulation.

At one level the executive brain's neocortex is a relatively simple anatomical structure, made up of six layers formed by radial migration of neurons from the basal cortical plate where active cell division takes place. Formation of cortical columns from the radially migrating neurons has provided the functional multicellular units of the neocortex. The number of cortical columns is dependent on the size of the cortical pool of stem cells that are produced early in development as a result of symmetrical cell divisions. Asymmetrical divisions of these neocortical stem cells are responsible for producing the number of cells in each column. The cells in each cortical column subsequently differentiate to form the six different layers of the neocortex. This period of cortical column neurogenesis varies across species, taking eight days in the mouse and extending to 80 days in humans for the production of each column. Although the development of new cortical column stem cells extends for longer with increasing cortex size, when such neurogenesis is expressed as a proportion of total cortical developmental time, there is a remarkable congruence across different mammalian species. This reinforces the viewpoint that there is strong conservation at the genetic level for the development of cortical columns. The major variance for this cortical column development across mammalian species occurs in the cellular resting phase. This takes place between the production of each new neuron during the period when it originates from the basal cortical plate. This developmental period becomes progressively longer with the increasing number of cell divisions, and hence with brain growth. Developmental timing is thus of great importance for neocortical executive brain development, and a progressive slowing down of this developmental process (called neotany) is thought to have contributed to the evolutionary increase in human brain size. It certainly takes longer, indeed longer on a logarithmic scale of increased time, for the human brain to develop compared with that of other primates. The frontal cortex in particular, undergoes very late reorganisational development, which occurs during the period of adolescence in humans. This region of the human cortex has a genetic transcriptome that is remodelled during this period, and is significantly delayed for its transcription in humans relative to all other primate species. Because it is the executive brain which plays the primary role in decision making for human reproductive and maternal behaviour, it is of considerable beneficial importance that these developmental decisions are made appropriately.

Nevertheless, it is not possible to develop a neural structure as complex in its interconnections as that of the executive brain without making mistakes, and such errors need to be accounted for during development. The magnitude of these errors is illustrated by a comparison of the developing neocortex in mouse and monkey. Taking into account the size of the starting pool of neurons in these two species, the number of divisions these cells undertake and the interval between cell divisions, the monkey cortex should theoretically achieve a size three times larger than is actually observed. It is also mathematically impossible, without incorporating a consideration for errors, to construct an algorithm that accurately accounts for neocortical brain development involving these billions of neurons and trillions of synapses. Evolutionary biology has resolved these vast developmental problems for brain developmental complexity by incorporating programmed cell death into its normal error-prone developmental programme. Cell death provides an epigenetic version of 'survival of the fittest' at the neuronal level, and has a strong element of stochasticity (Edelman, 2003). Only those neurons survive which are in the right place, reach the correct axonal targets at the right time and make the appropriate synaptic connections. From our knowledge of comparing the development of the mouse and monkey brains, programmed cell death is not trivial. Moreover, it is of no surprise that a number of the important imprinted genes are major regulators of neural cell-cycle arrest (preventing cells from multiplying) and, notably, they also regulate programmed cell death by apoptosis. These epigenetically imprinted genes and their complex networks have undoubtedly played a genetic and epigenetic leadership role in the evolutionary and developmental expansion of the cortex. We have evidence for the importance of epigenetic methylation in brain development from the failures of the genetic methylation enzyme, *MeCP2*, which results in severe autistic development. It is, of course, the genome's epigenetic methylation marks that also determine which genes are imprinted.

Timing, spatial relationships, directional cues, target cues and the growth rate of axons all require integration during cortical brain development. Moreover, all cortical neurons make contact with other neurons that undertake this very same process. Because the consolidation of neural interconnections is activity-dependent, the environment that generates this neural activity is clearly important. Thus no two brains are exactly the same and although genetic programmes do play an integral part in brain development, monozygotic genetically identical twins differ in many aspects of their behaviour and even in their predispositions to psychiatric disorders (Haque *et al.*, 2009). For these reasons it is clear that epigenetics also plays a very significant role in neocortical development, particularly taking into account those epigenetic imprint control regions that are germline reprogrammable, regulate gene dosage and are expressed according to parent of origin. Also important are those cortical 'escapee genes' which avoid demethylation reprogramming in the maternal germline (Tang *et al.*, 2016) (see Chapter 3).

Most primate executive neocortical development occurs postnatally and continues until the time of puberty. This development occurs in an environment dominated by matrilineal kin, thereby providing a socialising influence on their developing brain (Dunbar, 2003). A further reorganisation of the cortex, notably prefrontal and temporal cortical regions, occurs throughout puberty (see Chapter 4) and hence takes place in a social environment of developing sexual awareness. Executive brain function, and the way in which these regions of the brain regulate motivated behaviour, has resulted in much of such behaviour becoming emancipated from the deterministic action of hormones, especially in humans. Thus, we humans eat to avoid hunger rather than to satisfy hunger; most of our sexual activity is not reproductive; and maternal care can take place outside the context of pregnancy and parturition. The adolescent reorganisation of the executive brain, particularly the prefrontal cortex with its controlling connections to the emotional brain, thereby assumes an important regulatory control over the kind of behavioural decisions that humans make, especially in the context of primary motivated behaviours. The multitude of sexual proclivities that persist across and within different human societies may stem from the kind of sexually rewarding environment in which the prefrontal cortex undertakes this developmental reorganisation during adolescence. Some of the puberty-related maturational events of the neocortex are different between males and females (see Chapter 4), which may also contribute to differences in social and cognitive development during adolescence, and to the lasting sexual dimorphisms seen in the adult neocortex (Ryman *et al.*, 2014).

We can thus see how the evolutionary development of the executive brain has incorporated substantial epigenetic modifications which have enabled advantages of the kind that generate variability, and permit greater adaptability to those social environmental changes favoured by natural selection. However, such is the complexity of the human executive brain that this potential for variability in creating the brain's developmental interconnections may also be influenced by socially impoverished environments. Herein dwells the risk for brain development, a risk which is particularly well-illustrated when considering the potential for the broad spectrum of psychiatric disorders that human brains can experience.

REFERENCES

Broad, K. D. & Keverne, E. B. (2011). Placental protection of the fetal brain during short-term food deprivation. *Proc. Natl Acad. Sci. USA* 108: 15237–41.

Broad, K. D., Curley, J. P. & Keverne, E. B. (2006). Mother–infant bonding and the evolution of mammalian social relationships. *Phil. Trans. R. Soc. B* 361: 2199–214.

Cowley, M., Garfield, A. S., Madon-Simon, M., et al. (2014). Developmental programming mediated by complementary roles of imprinted *Grb10* in mother and pup. *PLoS Biol.* 12: e1001799.

Curley, J. P. & Keverne, E. B. (2005). Genes, brains and mammalian social bonds. *Trends Ecol. Evol.* 20: 561–67.

Dunbar, R. (2003). Psychology. Evolution of the social brain. *Science* 302: 1160–61.

Edelman, G. M. (2003). Naturalizing consciousness: a theoretical framework. *Proc. Natl Acad. Sci. USA* 100: 5520–24.

Eluvathingal, T. J., Chugani, H. T., Behen, M. E., et al. (2006). Abnormal brain connectivity in children after early severe socioemotional deprivation: a diffusion tensor imaging study. *Pediatrics* 117: 2093–100.

Haque, F. N., Gottesman, I. I. & Wong, A. H. (2009). Not really identical: epigenetic differences in monozygotic twins and implications for twin studies in psychiatry. *Am. J. Med. Genet. C Semin. Med. Genet.* 151C: 136–41.

Hinde, R. A., Leighton-Shapiro, M. E. & McGinnis, L. (1978). Effects of various types of separation experience on rhesus monkeys 5 months later. *J. Child Psychol. Psychiarty* 19: 199–211.

Keverne, E. B. (2006). Trophoblast regulation of maternal endocrine function and behaviour. In: A. Moffett, C. Loke & A. McLaren (eds.), *Biology and Pathology of Trophoblast*. New York, NY: Cambridge University Press, pp. 148–63.

(2013). Importance of genomic imprinting in the evolution and development of the maternal brain In: D. Pfaff & Y. Christen (eds.), *Multiple Origins of Sex Differences in Brain. Neuroendocrine Functions and their Pathologies*. Berlin: Springer-Verlag, pp. 21–34.

(2015). Genomic imprinting, action, and interaction of maternal and fetal genomes. *Proc. Natl Acad. Sci. USA* 112: 6834–40.

Keverne, E. B., Martel, F. L. & Nevison, C. M. (1996). Primate brain evolution: genetic and functional considerations. *Proc. R. Soc. Lond. B* 262: 689–96.

McCarthy, M. M. (2010). How it's made: organisational effects of hormones on the developing brain. *J. Neuroendocrinol.* 22: 736–42.

Pollard, S. K., Salama, S. R., Lamber, N., et al. (2006). An RNA gene expressed during cortical development evolved rapidly in humans. *Nature* 443: 167–72.

Ryman, S. G., van den Heuvel, M. P., Ueo, R. A., et al. (2014). Sex differences in the relationship between white matter connectivity and creativity. *Neuroimage* 101: 380–89.

Sato, Y., Shinka, Y., Sakamoto, K., et al. (2010). The male-determining gene *SRY* is a hybrid of *DGCR8* and *SOX3*, and is regulated by the transcription of CP2. *Mol. Cell. Biochem.* 337: 267–75.

Schultz, W. (2000). Multiple reward signals in the brain. *Nat. Rev. Neurosci.* 1: 199–207.

Striedter, G. F. (2005). *Principles of Brain Evolution*. Sunderland, MA: Sinauer Associates, Inc.

Tang, W. W., Kobayashi, T., Irie, N., et al. (2016). Specification and epigenetic programming of the human germ line. *Nat. Rev. Genet.* 17: 585–600.

Varrault, A., Gueydan, C., Delalbre, A., et al. (2006). *Zac1* regulates an imprinted gene network critically involved in the control of embryonic growth. *Dev. Cell* 11: 711–22.

The Epigenetic Landscape in the Evolutionary Ascent of the Matriline: Concluding Overview

Many biologists would consider the genetic origins for determining the mammalian male and female to reside on the Y- and the X-chromosomes. This represents the classical view of sex differences, and can lead to the kind of misconceptions that have arisen concerning females as representing the biological default genetic sex. Sex differences are not this simple, and such viewpoints are far from the reality of what we now understand about the genetics and, indeed, the epigenetics that underpin the development of mammalian sex differences. In this monograph I have tried to clarify the work relating to the biological status of the female and the evolutionary origins for the differences across and between the sexes. This thinking has engaged with a broad base of investigations that have been made across multiple levels of functioning. These investigations have included a consideration of developmental genetics, epigenetics, reprogramming of the genome, genomic imprinting, placentation, hormones, brain development and behaviour. Core to all of these levels of biological functioning, it is the female germline that has been of special importance in shaping the evolutionary success of mammalian life. The importance of this success has been most notably represented by the establishment of mammalian placentation which, together with expansion of the neocortex, has been instrumental in driving the reproductive success of eutherian mammals.

Major evolutionary adaptations have been necessary to successfully embrace these multiple levels of functioning, and for the female genome to engage with the male genome in the success of in-utero development. Although the evolution of viviparity and the maternal in-utero environment have supported the development of both sons and daughters equally, it is now becoming clear that the female genome, and in turn the female's physiology and behaviour, have taken a lead in managing the evolutionary burden of those changes that have made this possible. I know of no mammalian biology that supports considering the 'female' as contributing anything less than a major role in promoting the evolutionary success of mammals. At whatever level we consider female biology, be it starting at our genetic germ cell origins and finishing with the immense complexity of the human brain, we find it is the female genome/epigenome that has been integral to, and foremost in, ensuring the success of this evolutionary progression. Specifically, it is the matriline that has taken a lead for the many genetic and epigenetic developments that have given rise to the successful evolution of the sex differences as we recognise them today (Keverne, 2013). This female bias starts at the very earliest stages of the genome's contribution to development, playing an integral role in the reprogramming of the germ cells, in evolving genomic imprinting, and in sustaining masculinity throughout the gradual evolutionary erosion of the male's Y-chromosome genes (Cortez et al., 2014). Of course, the autosomal maternal and paternal genomes have a matching and necessary role in supporting the success of these developments, but from an evolutionary perspective, the selection pressures for change have operated primarily through the matriline. The question therefore arises as to the specific ways in which the matriline acquired such a primary role in the success of our human heritage, and what are the developments that have brought about this success.

As most biologists are aware, human males have an XY sex chromosome complement of genes and females have two copies of the X-chromosome genes. The male Y-chromosome at no time resides in the female while the male's X-chromosome is always inherited from the female. Two features set the male Y-chromosome apart from the rest of the genome. First, there is a lack of recombination (the process that leads to the formation of new genes and DNA repair) on the Y-chromosome over most of its length, and there is also a limited transmission of those few male genes still residing on this Y-chromosome. Apart from the male's important sex-determining *SRY* gene, the Y-chromosome has only 78 protein-coding genes compared with more than 2000 such genes on the female X-chromosome. Unlike the broad-based actions of X-chromosome genes, those on the Y-chromosome have been evolutionarily selected for male-only specific functions. Hence, the effectiveness of natural selection in acting on the evolution of the non-recombining Y-chromosome genes has become greatly reduced and, in its absence, natural selection treats the genes of the Y-chromosome as a single unit specialised for male functions. The X-chromosome houses a large number of genes which are expressed in the foetal placenta, and a second large group that are expressed in the developing 'emotional' brain. These two structures have a complexity of function that is greater than most of the body's other organs, especially considering the relatively few but unique cell types from which they are initially constructed (neurons and glia for the brain, trophectoderm for the placenta). Moreover, both the brain and the placenta have played a major role in the successful shaping of mammalian evolution. Early evolution of the placental mammals has employed imprinting for X-inactivation, always ensuring inactivation of the paternal X genes (Gribnau & Grootegoed, 2012). Although the evolutionary success of maternal X-chromosome genes has been of significance for both males and females, the matriline adopts a further leadership role in determining the deployment of maternal imprinting for many of the autosomal genes that are engaged in the construction of the brain and the placenta.

A major player in mammalian sex determination now appears to be the autosomal *Cbx2* gene. *Cbx2* is an autosomal gene which targets repressive chromatin complexes in a 'parent of origin' manner. The multiple copies of this gene in the male link to paternal pericentric heterochromatin, which epigenetically silences paternal gene expression. Mutations in the *Cbx2* gene result in a male to female sex reversal. Hence, *Cbx2* can be considered as a part of the pro-male pathway through the inhibition of *Foxl2*, the gene that is actively engaged in female ovarian development. In males the repressive *Cbx2* complex thus enables *Sox9* to be expressed. This gene (*Sox9*) is downstream of *SRY* expression, and is actively engaged in production of the testes. Hence, genetic development of the female ovaries first requires to be suppressed before the pro-male genetic pathway can be activated.

The male sex-determining *SRY* gene on the Y-chromosome is unique to placental mammals, but is not, as many textbooks would suggest, a completely new gene. *SRY* is in fact a hybrid of two pre-existing genes, *DGCR8* which is expressed in, and is important for, placental development, and *SOX3* which is important for developing those parts of the hypothalamic (emotional) brain which determine sexual behaviour in the male, and maternal behaviour in the female (Sato *et al.*, 2010). These areas of the hypothalamic brain are unique in their sexual dimorphism, a dimorphism which becomes established during early brain development, and is brought about by the masculinising action of the hormone, testosterone. Precision timing for *SRY* expression is therefore important for underpinning the developmental timing of the testosterone producing foetal Leydig cells. These too require to be developmentally synchronised with *Sox3* expression for determining male hypothalamic brain development, and thereby ensuring that hypothalamic development is synchronised to the restricted developmental window that is available for brain masculinisation. What better way could there be for ensuring the precision timing of masculinisation than to construct the *SRY* male sex-determining gene by hybridising component genes that are established for the development of (a) the foetal Leydig cells and (b) the foetal hypothalamus. Unlike the adult Leydig cells, foetal Leydig cells produce testosterone without the need for any intervening activation via the production of those hypothalamic- and pituitary-stimulating hormones which are required at puberty, and which continue to be necessary for the production of testosterone in the adult. At this early embryonic stage of development, it is the foetal Leydig cells which take the lead for production of testosterone, and subsequently they undergo apoptosis (cell death) in the early post-partum period. For this reason, in-utero masculinisation of the male's brain is integral to determining adult male sexual behaviour. Such masculinisation also diverts the developing male hypothalamus away from a maternal developmental trajectory, which is also brought about during the in-utero period. In this case, it is the hormones originating from the foetal placenta that play the primary role. Importantly, the male *SRY* gene is not functional in determining male sexual differentiation without the receptor for testosterone. The gene for this receptor is expressed from that X-chromosome which males always inherit from the mother, again providing the matriline with ultimate developmental control over the future sexual and parental behavioural differences between males and females.

In female mammals, one of her two X-chromosomes becomes epigenetically silenced, thereby ensuring that equivalent levels of female X-gene dosage match to male X-gene dosage. Males have only the one X chromosome which they always inherit from their mother, while gene silencing for one of the female X chromosomes, notably the one she inherited from the father, is established at the earliest stages of development (Berletch et al., 2015). Female X-chromosome silencing is brought about by the production of a long non-coding RNA, namely the X-inactive specific transcript (Xist) (Gendrel & Heard, 2014). This long non-coding RNA (Lnc-RNA) coats the future inactive X-chromosome, thereby triggering an epigenetic cascade of chromatin changes and DNA methylation. In this way, a lifetime of repression for genes on one of the female X-chromosomes is accomplished, a repression that was originally thought to be random, but studies in mice have now shown X-inactivation to be directed only to the paternal X-linked genes that are expressed in females. The Xist non-coding RNA transcript triggers gene silencing in *cis*, together with a gain of repressive histone chromatin marks, and accompanied by a loss of those histone marks that are permissive for X-gene expression. Subsequently the inactive state of the X-chromosome genes are locked in place by methylation of DNA at gene promoters (Prudhomme & Morey, 2016).

During the earliest cell divisions of the fertilised egg (four-cell stage) the first phase of X-inactivation specifically targets the paternal-X. This specific early silencing of the paternal X-chromosome is thought to be brought about by a germinal mark laid down on the maternal X-chromosome during oogenesis, thereby protecting the female X-chromosome from silencing. Such imprinting of X-gene inactivation for silencing of the paternal X-chromosome ensures these paternal alleles remain silent in the trophectoderm cells which proceed to develop into the foetal placenta (Merzouk et al., 2014), and thereby avoid potential rejection of the foetus by the mother's immune system.

Whereas the Xist RNA inactivates only one of the X-chromosomes of females, such inactivation needs to be avoided in males because they possess only the single X-chromosome, namely the chromosome which they inherited from their mother. It is the unpaired hemizygous state of the Xist chromosomal region in males which prevents inactivation of this single X-chromosome which they inherited from the mother (Sun et al., 2015). Hence, males always inherit and express genes from the maternally inherited X-chromosome, while females show imprinting of the X-inactivation process, and thereby silence most of the genes they inherit on the paternal-X. In this way, the female germline controls the destiny of those X-chromosome genes which they inherit from the male. The female germline also ensures that the maternal X-chromosome, which females pass on for expression to the next generation of males, has first been stabilised by DNA mismatch repair, again by the matri-line. Such mismatch repair of the single male X-chromosome genes has no matching female copy and cannot undertake this repair process in the male germline. However, because it is always inherited from the female, this is not such a problem, as more errors are likely to occur due to the multiple cell divisions which take place during spermatogenesis. The female germline can thus be considered as the intergenerational repair station for male X-chromosome genes.

In summary, early mammalian development witnesses a leading role for the matriline in the stability and heritability of X-chromosome gene expression. This ensures that it is always the genes inherited from the maternal-X which are expressed in the male placenta and, through maternal genomic imprinting, those X-genes that females inherit on their paternal chromosome are silenced in the placenta of the female foetus. The pairing and passage of X-linked genes through the matriline for both sons and daughters also ensures they undertake mismatch repair, thereby providing genetic stability for development of sons as well as for daughters.

GERMLINE REPROGRAMMING

Gaining a clear understanding of the biological events upon which natural selection has operated for mammalian evolution requires a much broader context of enquiry than that taken in the past. Thus, in humans, and mammals in general, it is the female oocyte that commences the next generation's reproductive life, ahead of the male's genetic contribution to the next generation. The female produces larger gametes, and produces them before birth during the period when their female owner is still in-utero. It is during this earliest stage of in-utero development that the female germline carries the primary responsibility for sustaining the genetic viability of the next generation. This they achieve by the epigenetic reprogramming of their genome to a totipotent state in readiness for developmental conception of their next generation. Moreover, following fertilisation it is again the matriline which engages with reprogramming of the male epigenome, as well as further undertaking mismatch repair of the sperm's DNA. Not only does this bringing together of male and female germ cells give rise to the next generation, but as a consequence of their capacity to undertake demethylation reprogramming, this process of restoring totipotency to the germ cells provides the foundations for an endless series of future 'like' generations.

At the very earliest zygotic stages in development of the mammalian fertilised egg, DNA methylation profiles can be found that are specific to each of the cell lineages that provide either for the origins for the placenta, or are destined to form the embryo itself. Those differentially methylated regions of the genome that are found in the earliest blastocyst cells are initially determined by the more flexible methylation of histones. This functions as a 'landmark' that is responsible for setting up the DNA methylation profiles for future specific cell lineages. In the context of placental development, these very early genetic landmarks are themselves derived from the repeat sequences located in retro-viral DNA, suggesting these foreign DNA sequences may have helped to drive cell fate in the evolution of placental mammals (Wu et al., 2016).

Although all vertebrate life starts from a single oocyte and spermatozoa which unite to form the fertilised egg, the fundamentals of maternal and paternal germline reprogramming has itself changed over time. As vertebrates evolved, so too have there been substantial modifications to the evolutionary process for germ cell reprogramming. Such changes further illustrate the growing importance of the matriline in mammalian evolution, changes which have impacted on the fundamental process of generating mammalian viviparity and maternal caring for offspring, both of which are essential events which embrace the continuity of mammalian life.

If we first consider fish as an example of early vertebrate evolution, it is here that the male takes the leading role for reproduction, sometimes building and guarding the nest, and competing with other males for the female's attention. Both female and male fish produce an order of magnitude more gametes than do mammals and, from an evolutionary perspective, their reproductive success can tolerate a considerable number of embryo failures. Integral to understanding such different vertebrate reproductive strategies at the genomic level requires further knowledge of the very earliest stages of gamete production. Indeed, all vertebrate life, including that of humans, begins by turning back the intergenerational clock for germline gene expression to ground zero (epigenetic reprogramming). This is followed by the subsequent bringing together of male and female gametes at fertilisation. Reprogramming of the genome is fundamental to achieving totipotency for the genes of this next generation. Although few species have been studied so far, major differences have already been found between non-mammalian vertebrates and present-day mammals. These differences provide us with evolutionary insights into how epigenetic reprogramming of the genome has progressed. In the egg-laying vertebrate fish, reprogramming is a male-dominated regulatory control mechanism (Murphy & Cairns, 2016), in contrast to that encompassing a matrilineal lead for epigenetic reprogramming in the viviparous mammals.

The zebrafish is one of the few non-mammalian vertebrates for which germline reprogramming has been studied in detail, and reveals the maternal genome of their oocytes to be profoundly silenced (methylated) and unavailable for transcription. This is in marked contrast to the hypomethylated, and hence available, status of male zebrafish sperm DNA. On fertilisation it is the paternally derived epigenetic state of the fish genome which attains heritable stability in this non-mammalian vertebrate (Murphy & Cairns, 2016). Moreover, it is the post-fertilisation genome of the female fish oocyte which undertakes extensive remodelling, thereby resulting in the epigenetic state of the maternal genome matching to that of the paternal genome (Potok et al., 2013). Thus, the genes that are important for early development in this vertebrate are paternal, while the many oocyte genes of the female are methylation-silenced and require reactivation by reversal of their methylation status in order to function. Hence, early fertilised eggs of the zebrafish achieve a totipotent state through paternal competence taking the lead, and the maternal genome undergoing epigenetic reprogramming to match that of the father (Hackett & Surani, 2013).

In mammals there are two stages to reprogramming the genome of the male germline. Initially this occurs in the early phases of neonatal sperm development and involves those stem cells that will eventually support continuous sperm production in the adult. The second phase of reprogramming for the male genome occurs after fertilisation, and this is primarily under the control of the female's (oocyte) genome (Leseva *et al.*, 2015). The earliest initial phase of mammalian male genome reprogramming is primarily one of DNA methylation, ensuring silencing of the male genome as a prelude to DNA condensation and DNA repackaging by sperm-specific proteins (protamines). These protamines compact the mammalian male genome to a condensed structure, reduced in size to occupy less than 10% of the sperm nucleus. High levels of DNA compaction are thought to optimise sperm shape and thereby enhance the sperm's swimming capacity for its long journey to the oocyte. The reversal of such dense DNA compaction for the male sperm thus requires maternal assistance after fertilisation in order for post-fertilisation reprogramming of this silenced mammalian male genome to take place. Interestingly, the male zebra fish sperm genome is free from any compact protamine packaging and readily available for transcribing.

Such a fundamental shift from a paternal to a maternal lead in the evolutionary progression of germline reprogramming is only the beginning of this complex reprogramming story. The active role played by the matriline has seen further evolutionary changes at the genetic level within mammals. In the early development of mice, a tripartite genetic network acts in combination to repress somatic genes, induce germ-cell gene expression, and re-induce the expression of those genes that initiate the genetic programme for development. The *Sox2* gene is key to this early mammalian (mouse) germline reprogramming (Hackett *et al.*, 2013). In humans, the very different *Sox17* gene is the key regulator for specification of the primordial germ cells, and *Sox2* is repressed in human germ cells (Tang *et al.*, 2016). Associated with *Sox17* is the repression of another gene, *Blimp1* in humans, and this functions to repress gene expression in somatic cells. Thus, at this earliest and most fundamental level of developmental reprogramming in the viviparous mammals, genetic divergence has occurred not only from their ancestral vertebrates, but also within the mammalian lineage itself.

Across the placental mammals, the very earliest stages of development are under active selection pressures for differentiation of those numerous tissue types that are determined by early gene expression. Interestingly, some of these developmental genes escape early demethylation and thus are not available for expression. Such genes include those which are subsequently expressed in the brain and have developmental links to neurological disorders (schizophrenia) and metabolic disorders (obesity) (Tang *et al.* 2016). It is still early days in our knowledge of these 'escapee from de-methylation genes', but presumably such brain pathologies represent the failure of these genes to regain functional expression should they continue to remain methylated and silenced during the brain's extended development.

The erasure of epigenetic gene modifications (demethylation of histones and DNA) in the germline genes of the pre-implantation embryo, has served as an important barrier to epigenetic inheritance. However, the finding of 'escapee genes' that are resistant to demethylation during the period of early germline reprogramming may provide potential candidates for transgenerational epigenetic silencing. We might ask the question as to what advantages such escapee genes provide for brain development and brain evolution, since many of these genes are expressed in the brain. Interestingly, the developing brain expresses its own demethylating enzyme, thereby providing a germline independent mechanism for reversing gene silencing. In this way, escapee genes may have become conserved for, and thereby subject to, later selection pressures that are specific to brain evolution. We might therefore ask the question as to the advantages which 'escapee' genes provide, since many of them are subsequently expressed in the brain. Interestingly, the brain expresses its own de-methylation enzyme which provides a germline independent mechanism for reversing gene silencing. This ensures these genes are functionally available for those later brain developments which occur at puberty, and are thereby conserved for selection pressures that operate during this pubertal period. Puberty onset is very responsive to environmental events and escapee genes may thus play an important role in this response, a role which is under the plastic epigenetic regulatory control of brain function, rather than early germline reprogramming. However, a failure to undergo later demethylation may, in some cases, be responsible for the subsequent development of schizophrenia, and metabolic disorders. 'Escapee' genes are found in both mouse and human, although many more of these escapee genes are present and subsequently expressed in the human brain.

DNA REPAIR

The successful achievement of mammalian in-utero development places a heavy burden on the mother in achieving success for this mode of reproduction, and with it comes a high risk for the mother herself should errors in foetal development occur. Not only could developmental genetic errors potentially result in loss of the foetus, but such early foetal loss carries with it a high risk for the mother's future reproduction, and even the possibility of maternal fatality. It therefore stands to reason that, with this disproportionate burden of risk focussed on the mother, natural selection has provided the maternal genome with a leading role for the repair of foetal DNA during the regulatory control of early development. Indeed, genome-wide reprogramming in the mouse post-fertilisation germline entails deployment of the base excision repair pathway (Hajkova et al., 2010).

Following fertilisation but prior to implantation, DNA inherited from the mother and from the father reveal a very different expression status. Considering the paternal mitochondrial DNA, which is the source of energy for sperm motility, then this becomes energetically exhausted from the sperm's marathon swim to reach the oocyte, and fails to gain access to the female oocyte. Only maternal mitochondrial genes are passed on to the next generation. Moreover, sperm nuclear DNA is characterised by point mutations, extended tandem repeats and a predominance of structural chromosomal mutations (Marchetti et al., 2007). These result from the large number of cell divisions that are required for the process of successful spermatogenesis. Thus, following fertilisation, it is the father's nuclear chromatin that becomes remodelled, and at the nuclear replication phase, the sperm's DNA undergoes a second phase of demethylation linked to repair of any mismatches with female DNA (the mother's oocyte) (Santos *et al.*, 2013). Hence, DNA repair at the early zygotic phase of cellular genome reprogramming can be regarded as a maternal trait for repairing paternally derived errors. Reprogramming of paternal DNA is also aided by the inheritance of several maternal factors from the oocyte. Thus, maternally expressed genetic factors bring about demethylation of the paternal genome (*Gse*) (Hatanake *et al.*, 2013). Other maternally expressed genes are required to protect and maintain maternal imprints (*Pgc7*, *Zfp57*, *Stella*) during this post-fertilisation phase of paternal demethylation (Li *et al.*, 2008). The maternal genome also initiates specific differentiation of the cells that go to form the future placenta (*Cdx2*, for trophectoderm cells). Also following fertilisation, the maternal genome aids removal of the sperm's compact protamine packaging, a packaging that tightly maintains silencing of paternal DNA. These paternal packaging protamines are replaced with histones, many of maternal origin, as a necessary requirement for paternal DNA to commence transcription (Yang *et al.*, 2015). Thus, the contribution of paternal DNA to early foetal embryonic development is carefully chaperoned by maternal DNA to avoid errors during foetal development, errors that could be a risk to the mother's reproductive success and to the mother herself.

GENOMIC IMPRINTING

Genomic imprinting is unique to placental mammals and represents a matrilineal epigenetic innovation that has played a significant role in the progression of mammalian evolution (Ferguson-Smith, 2011). The imprint control regions (ICRs) are primarily determined by maternal DNA, and are responsible for the regulation of autosomal gene expression according to its parent of origin. That is to say, some autosomal genes are only expressed when originating from the mother, and others are only expressed when originating from the father. The maternal genome takes a further leading role in the post-fertilisation reprogramming of these imprints for the following generation, while the genes which are imprinted have themselves undergone 'purifying selection' (Hutter et al., 2010). Purifying selection renders these genes free from mutations and very stable across generations, an important and necessary requirement for the success of in-utero foetal development. Thus imprinted genes differ in expression according to their parent of origin, and although their imprint control regions arise in, and are inherited primarily through the matriline, these genes are expressed in both sons and daughters.

In-utero development for the offspring has undoubtedly placed the burden for successful reproduction firmly under matrilineal control. Therefore, it is not surprising that genomic imprinting first appeared in the earliest of placental mammals, the marsupials. There is no evidence for genomic imprinting in the egg-laying monotreme mammals and there are relatively few (six identified so far) imprinted genes engaged in the development of the short-lived yolk-sac placenta of the marsupial wallaby (Renfree et al., 2013). The number of imprinted genes has increased exponentially during the evolution from marsupials to eutherian placental mammals. Some150 genes have been identified that are imprinted, show monoallelic expression and involve multiple tissues. Many additional genes that are tissue specifically imprinted, have also been identified to date. The latter are often referred to as having a tissue-specific allelic bias for expression.

It therefore seems likely that, as placental complexity evolved, and in-utero development proceeded for a longer period of time, the need for tighter control over gene networks resulted in the imprinting of more genes than those initially identified and expressed in the short-lived metatherian placenta. Imprinted genes are not themselves necessarily new genes, but established genes that have migrated to the ICR, enabling them to be secured for monoallelic expression and purifying evolutionary selection through DNA repair mechanisms. Many of the imprinted genes that are expressed in mammals are integral to the successful development of the placenta, and are co-regulated as part of an imprinted network (Varrault et al., 2006). Moreover, each imprinted gene may itself regulate a downstream network of genes engaged with specific functions.The stability of imprinted genes and the robustness of their functional networks thus allows for interacting genetic hubs to provide compensatory actions through their common downstream genes, should this be needed (Sandhu, 2010). In this way, any minor errors which may occur in gene transcription can be developmentally compensated for across the expressed genomic network.

Operating via the foetal placenta, the imprinted genome ensures transfer of nutrients to the foetus, and the production of hormones that instruct the mother to provide the protective warmth of a nest before the foetus is born. Most importantly, it is these foetal placental hormones which drive the mother to greater food consumption in advance of the substantial energetic demands that will subsequently arise as pregnancy progresses and the foetus develops. Furthermore, these placental hormones ensure that the physiological priorities of the pregnant mother are also focussed on the developing foetus, thereby shutting down the mother's reproduction and her sexual activity throughout both pregnancy and during the lactation period of most mammals (Keverne, 2006).

Genomic imprinting has had considerable impact on mammalian development and evolutionary success, primarily through regulation of the brain's behaviour, foetal placentation and mammary gland production of milk. The imprinted genes themselves are remarkably stable as a result of purifying selection, while their epigenetic ICRs are responsible for gene regulation and haploid production of gene dosage. The maternal epigenetic marks for imprinting are produced by DNA methylation. This methylation mark represents the primary maternal imprint that results in paternal gene expression. However, as these imprints also regulate inhibitory non-coding RNAs in the imprinted gene cluster, they are also instrumental in permitting only maternal allelic expression. This is also the case for retrotransposed elements that are associated with the maternal imprints and which have provided the mechanism for silencing paternal gene expression.

Genomic imprinting has gained in complexity from the networks they form with other imprinted genes (Varrault *et al.*, 2006). This has resulted in a further broadening of their functional impact over both developmental time and space expression, coordinating stable gene dosage across different tissues during development. For example, the paternally expressed gene (*Peg3*) has a network of 22 genes concerned with neural development and 21 genes that regulate other transcription factors concerned with metabolism, body temperature and feeding, the dysfunctioning of which leads to obesity and diabetes (Curley *et al.*, 2004). Another imprinted gene cluster which is maternally expressed, and particularly affects the brain, is the Angelman's Syndrome cluster, named after its discoverer, and which is now known to be a consequence of deficiency for the maternal *UBE3a* gene (Meng *et al.*, 2015). This syndrome is characterised by intellectual dysfunction, speech impairment, seizures and ataxia. The paternal copy of this imprinted gene is normally not available for expression due to its silencing by the maternal non-coding RNA gene, the *UBE3a* anti-sense transcript.

As previously mentioned, the large brain of primates, and especially the human brain, are exceptional in the context of mammalian behaviour. In such large-brained primates, sexual behaviour has, to a notable extent, become emancipated from the deterministic actions of hormones on the brain. The same is true for maternal behaviour, where the need for pregnancy hormones is not essential to ensure maternal care and mothering attention. Indeed, it was not so long ago that the aristocracy in England employed 'wet nurses' to nurture and even to breastfeed their babies. The possibility and the success of such alloparenting has depended on the massive evolutionary expansion of the human executive brain which has enabled human maternal, as well as sexual behaviour, to become emancipated from the determining action of hormonal regulation (Keverne, 2014). Nevertheless, just like all other mammals, the human female organs that function for milk production (mammary glands) and for reproductive fertility (ovaries) are still maintained by these phylogenetically ancient and reliable regulatory hormonal mechanisms during pregnancy. Moreover, the enhancement of genetic stability at this level of mammalian viviparity has resulted from the maternal imprinting of genes, a process that has also been essential for the successful transgenerational cross-talk between the developing foetal placenta and the developing mother's brain.

MOTHER'S BRAIN AND THE FOETAL PLACENTA: AN INTERGENERATIONAL TEMPLATE FACILITATING EVOLUTIONARY CO-ADAPTATION

In-utero development places the burden for reproductive success firmly on the mother, primarily through the co-adaptations that have occurred intergenerationally between the mother and her foetus. Such co-adaptations have been integral to the success of in-utero development, and especially for placentation. The placenta is genetically of foetal origins, but its development and functioning has been successfully brought about by co-adaptation with the mother's genome at a number of levels. Interestingly, many of those genes which have become co-adapted also take their origins for expression from matrilineal imprinting control mechanisms (Keverne, 2015).

One priority for successful in-utero development has been the prevention of maternal immunological rejection of the foetus. Moreover, it is not just the mother's immune system that has needed to co-adapt to the presence of a placenta, but the mother's brain has also required functional co-adaptation. This has enabled the mother to respond appropriately to the intergenerational hormonal signals secreted by the foetal placenta. These hormones instruct the mother to consume more food in advance of the energetic demands of her conceptus, to build a nest for protective warmth for the future offspring, and to inactivate further reproductive behaviour while the mother is pregnant and nursing. Finally, the foetal placental hormones prime the mother's brain for maternal care and prime her mammary glands for milk production after birth. Here, too, postnatal infant suckling further maintains the inactive silencing of mother's reproduction, thereby ensuring that this infant, or this litter in the case of small-brained mammals, receives the mother's undivided attention. All of these important maternal events are initially regulated by the foetal genome directing changes to hormone production by the foetal placenta, an intergenerational co-activation that is continued postnatally by the infant's suckling behaviour (Keverne, 2013). These events are a clear illustration of intergenerational foetal signalling to direct appropriate functioning of the mother's brain. The question arises as to how the maternal female brain evolved to read and respond appropriately to these foetal placental signals. The mother's brain is, after all, a generation ahead of the foetus. This intergenerational epigenetic co-adaptation must itself have become heritable as a consequence of genomic imprinting, as specifically directed mutations to such imprinted genes severely impair this mother–infant co-adaptive process (Keverne, 2015).

The mammalian in-vivo strategy for successful development involves both a maternal and foetal genetic commitment to the placental cell lineage at the very earliest stage of zygotic development. The maternally expressed gene (*Cdx2*), which initiates development of the cells that go to form the future placenta (trophectoderm), is one of the earliest genes to be expressed after fertilisation (Jedrusik *et al.*, 2015). Maternal factors are also responsible for the release of paternal DNA from its protamine cage following fertilisation, and thereby also making it available for transcription. This is in contrast to the sperm DNA of non-mammalian vertebrates which have no protamine packaging, thereby making their paternal DNA immediately available for transcription at fertilisation. Thus evolution has provided a fundamental epigenetic asymmetry between parents and offspring, consistent with the paternal genome achieving a competent state for early developmental gene expression in non-mammalian vertebrates (zebrafish), while it is the matrilineal genome that has been essential for enabling control of this same, early competent state for gene expression in mammals.

Intrauterine foetal development has itself introduced a new intergenerational dimension to understanding the evolutionary selection pressures acting on early mammalian brain development. Those parts of the developing hypothalamic brain of the foetus, which subsequently become responsive and active in determining multiple aspects of adult maternal behaviour, actually develop in-utero at a time when the foetal placental hormones engage with these same mature brain structures of the previous adult generation (mother). Successful foetal hormonal engagement with the mother's brain thus requires intergenerational epigenetic co-adaptation between this mother's brain and the foetal placenta, which produces the hormones that are provided by the developing next generation. It is therefore clear that the female brain cannot be a default from masculinity. On the contrary, among viviparous mammals, the genetic and epigenetic shaping of masculinity has became a positive and necessary development to avoid becoming maternal (see Chapter 6).

The steroid hormones of the foetal placenta are thus able to regulate much of maternal physiology and behaviour during pregnancy. Thereby, they ensure maternal resources for both successful placental development, and for that of the foetus. Foetus and placenta both arise from the same fertilised egg. Indeed, the earliest genetic decision of the fertilised egg, be it destined to become male or female, is to make a placenta long before any recognisable foetus is formed and even before the male genome is itself activated. As already mentioned, it is the maternal allele of the *Cdx2* gene which initiates placental trophectoderm formation. Further transgenerational interactions extend into the postnatal period, but these mother–infant interactions result from the advanced priming of the mother's brain, again by foetal placental hormones. In this way, it is the placenta that ensures there is adequate maternal milk supply available for its next generation, together with a warm nest constructively provided via the mother's brain, a brain which has in turn been primed by hormones from the foetal placenta for inducing all aspects of maternal care.

An important question arises as to how the genome for the next generation's foetal placenta has evolved the optimal way for regulating the adult maternal brain (hypothalamus) of its previous generation. Equally perplexing is the question as to how the adult maternal brain (hypothalamus) has itself developed to appropriately respond to the future demands of the following generation's developing placenta. These coordinated intergenerational events have each required co-adaptations across maternal and foetal genomes and epigenomes, the success of which is carried forward to the next generation. Notably, this is where the mammalian unique and maternally epigenetically imprinted set of genes are very important. A subset of these imprinted genes are coexpressed in the developing hypothalamus and placenta (Curley et al., 2004). Moreover, inserted mutations to these genes specific to the mother's hypothalamus, or independently in the developing foetal placenta, each severely impair maternal behaviour and the provisioning of milk. Such coexpression of the same imprinted allele in both structures (placenta and developing hypothalamus) thereby provides a unitary template on which natural selection may operate for co-adaptation. A number of these coexpressed imprinted gene clusters contain non-coding RNAs which are regulated by the *Drosha/Dgcr8* complex which is also coexpressed in the developing brain and placenta (Buckberry et al., 2014). Recent studies have provided evidence that non-coding RNAs are involved in brain development and, if dysfunctional, may result in both brain psychiatric disorders as well as dysfunctions of the placenta (Cheng et al., 2014). The imprinted C19 cluster of micro-RNAs has been shown to be associated with such pregnancy complications. Hence, through genomic imprinting, the matriline again takes the epigenetic lead for in-utero development, a lead that is beneficial to the successful development of both sons and daughters.

There are a number of other questions that arise from the invasive nature of in-utero development in mammals. Perhaps the most fundamental question concerns the immune system, and why the deeply invasive foetal placenta is not rejected, as is indeed the case for an organ transplant if the recipient's immune system is not suppressed (Loke, 2013). With respect to the foetal placenta, this is a complex immunological story, but clearly a global suppression of the mother's immune system cannot be the answer. This would create severe problems for avoiding infection during pregnancy. There are two possible ways around this problem: either to avoid the invasive placenta being seen by the mother's immune system, or to be seen but to localise active suppression of the immune response specifically to the maternal–placental interface. The foetal placenta engages both of these tasks but in different contexts, and again it is the maternal genome of the placenta which takes the lead in accomplishing these functions.

Considering the 'avoid to be seen' scenario for avoiding immune rejection, then at the very earliest stages of post-fertilisation embryo development, it is notable that the paternal genome of the zygote is silent, locked in its protamine cage, awaiting activation by female maternal factors. The outer layer of the embryonic blastocyst which forms the trophectoderm (future placenta) is determined by the maternally expressed gene *Cdx2* (Jedrusik *et al.*, 2015). Subsequently, the outermost layer of the developing placenta (syncytiotrophoblast) invades the mother's uterus, and engages with the mother's vascular system both for access to nutrients and for transferring placental hormonal messengers into the mother's bloodstream. Why then is the placenta not rejected by the mother's immune system as being alien? The answer is that the syncytiotrophoblast cells are devoid of the molecules (HLA epitopes) that induce immune rejection (Colucci *et al.*, 2011). The outer cell layer of the syncytiotrophoblast is immunologically invisible to the mother's immune system. This surface layer of the placenta contains molecules (syncytins) that are produced by a retrovirus (*HERV* genes), which has the ability to fuse cells, transforming them into a syncytial sheet (Muir *et al.*, 2006). These retroviral genes also have immune-suppressive properties and produce a protein, pleotropin, which acts as a growth-promoting factor for the placenta (Ball *et al.*, 2009). A third immunological factor influencing the life history of the placenta is the maternal NK (natural killer) cell, which infiltrates the area around the uterine-implanting placenta. In fact, these maternal NK cells are able to distinguish the HLA epitopes of the placenta and secrete growth factors which actually promote the development and growth of the placental blood vessels (Loke & King, 2000). Hence, mechanisms have evolved in the developing foetal placenta for expressing only the mother's alleles in the early trophectoderm cells, and for the subsequent development of a retroviral-driven immunologically invisible surface syncytial layer. Moreover, it is the mother's immune system (uterine NK cells) which prevents maternal rejection and actually facilitates placental growth (Karre, 2008). These positive in-utero co-adaptive intergenerational interactions between the mother and her foetus finally close the book on the view that the female represents a reproductive default state of the male.

EPIGENETICS, BRAIN DEVELOPMENT AND BRAIN EVOLUTION

The mammalian brain did not evolve as a de-novo unitary structure, but inherited a considerable past history from its long residence in the ancient vertebrates. This history is most notably represented in modern-day mammals by the structure and function of their limbic emotional brain. These ancient limbic regions of the brain represent the earliest phase of mammalian brain evolution, a phase which has been underpinned by a new way of mammalian gene regulation, namely genomic imprinting. Under maternal control, genomic imprinting has provided developmental guidance and genomic stability as a result of the monoallelic expression of certain developmental genes according to their parent of origin.

Those limbic regions which constitute the emotional brain (the hypothalamus in particular) are remarkably similar in structure and function across most mammalian species, and control the fundamental aspects of life that have been essential to the survival of all mammals. Different regions of the emotional brain are tightly linked to basic bodily functions (feeding, drinking, maintaining body temperature, reproduction and maternal behaviour) and respond to the hormonal messengers from the endocrine glands and, additionally in females, from the placenta of the next generation. However, substantial evolutionary changes to the functioning of these regions of the brain have taken place in the context of viviparity, most notably as a consequence of the progressive enlargement of the executive neocortex. These executive regions of the brain have played a major role in the brain's decision-making process, taking over control for some of the basic bodily functions, especially in the context of behaviour. Executive brain development has also required a tightening up of gene dosage by monoallelic gene expression together with a means of eliminating any developmental errors by employing programmed cell death.

The later phase of brain development, indeed the greater part of mammalian brain evolution, has occurred within the neocortex (executive brain). Most neocortical brain development takes place postnatally, extending into the period of adolescence in humans when a developmental reorganisation of these areas of the brain is brought about. This extensive period of developmental reorganisation is not complete for the human brain until adolescence is complete, and has enabled a logarithmic developmental increase in the size of the mammalian neocortex prior to undertaking reproduction. Moreover, this neocortical size increase is even greater across mammalian lineages than that which is seen between early mammals and their reptilian ancestors. The developmental growth of the mammalian neocortex reaches a peak in the primates and particularly in hominids. Primates are recognised for the exceptional complexity in their social interactions. Their social lifestyle has provided a greater need for the regulatory control of those motivated behaviours which are determined by the emotional brain. Thus, in large-brained primates, including humans, parenting is not restricted to the post-partum period, and sexual interactions are not restricted to the period of fertility. Feeding is undertaken to avoid hunger, rather than being determined by hunger. Such important evolutionary developments in the regulatory control of motivated behaviour have been made possible as a result of the evolutionary growth expansion of the neocortex and the important role this plays in the rationalisation of behavioural decision taking. Such progressive changes in the brain for the decision-making process have come about, at least in part, with the emancipation of behaviour from the determining action of hormones, especially in such large-brained primates. Moreover, the fact that much of neocortical development takes place postnatally, during a period of close relationships with the mother and subsequently with other members of the social group, has also been considered to provide an important selection factor in the evolution of the primate's executive brain (Broad *et al.*, 2006). Social living has been integral to the evolutionary success of primates, and has provided an environment that is of special importance for epigenetically shaping the development of the human executive brain. Once again, we find it is the matriline which takes on the leading role in this socialisation process.

Among the highly social monkeys it is the females, rather than the males, who determine the stability and cohesion of social groups. This is maintained over successive generations, with the social rank of daughters, but not sons, being epigenetically inherited from mothers (Dunbar, 2003). Thus, it is again the matriline that has provided the pivotal role in social evolution, and the neural mechanisms subserving this social role engage those same regions of the brain that are also involved with mother–infant bonding. These important neural circuits are linked to the biology of brain reward (Schultz, 2016). Moreover, the affiliative behaviour underpinning brain reward, such as grooming and sexual behaviour, release the same neuropeptides in the brain (oxytocin and beta-endorphin) that are also released when the mother gives birth (Curley & Keverne, 2005). Having developed the distinctive function of rewarding social encounters, these neuropeptides have subsequently acquired a defining role as the 'social glue' in the formation of complex societies. This is primarily underpinned by the developmental role of these same neuropeptides as part of well-being, and has been provided by the mechanisms originating in mother–infant bonding.

With the evolutionary expansion of the neocortex, and in turn the emancipation of behaviour from the determining action of hormones, maternal care is also available to, and indeed undertaken by, sisters and maternal relatives, without themselves undertaking pregnancy. The mother is thus not required to delay her next pregnancy until the brain of her offspring is fully mature. The extended matriline fulfils this long-term guidance through caregiving and, surveyed by the watchful eye of the matriarch, older daughters and younger sisters also gain mothering skills before undertaking pregnancy themselves (Keverne, 2004).

Such evolutionary developments raise the question as to what it is about the human brain that enables males to participate in parental care, albeit less consistent and less fulfilling than that of mother herself. The lack of emotional fulfilment for the male we know to be due to the physiological absence of parturition, the lack of functional male mammary glands, and the absence of suckling by the foetus. These physiological events, which are characteristic of mammalian mothering, are absent in the life of males. This, together with the lack of the emotionally reward-stimulating hormones, which the placenta provides to the female brain, may account for why the male brain is less predisposed to mothering. Nevertheless, the answer to the question as to why it is that human males are still able to exhibit some parental participation really resides in the extensive developmental evolution of the neocortex. This has elevated human behaviour to a higher plain, one which is strongly dependent on learning, memory and rationalisation. These are all events which the neocortex can readily undertake without the need for hormonal determinants. In line with this, a region of the human neocortex has expanded way beyond that of other mammals. This is the prefrontal cortex, which is involved with the temporal organisation of behaviour (Goldman-Rakic, 1987). Put in simple terms, the prefrontal cortex supports cognitive functions that are an integral part of social behaviour, and requires social knowledge that is organised both over time and in complex contexts (Kolb et al., 2012). Even in those mammals with a relatively small cortex (rats and mice), it has been shown that early behavioural experience dramatically alters the neural circuitry in the frontal cortex and the subsequent behaviour of these animals. However, such behaviour in laboratory rats and mice is certainly less complicated at the neural level, primarily because it is strictly tuned to rationalising the relatively discrete cues provided by chemosensory information (pheromones and olfactory cues). Olfaction is a unidimensional sensory system, processed by the allocortex (trilaminar cortex) in contrast to the more complex six layers of the neocortex that process behavioural interactions. Moreover, the social information carried in olfactory cues can be perceived even in the absence of behavioural interactions. These cues relate to the reproductive condition of females and to the dominance status of males. Such information is of particular importance to those many mammals that are nocturnal and live underground during the daytime.

At the cellular level, the neocortex of mouse and human is made up of similar types of neural cells, including cortical projection neurons, cortical reception neurons and GABA-ergic interneurons. These neurons collectively make up the basic units of the cerebral cortex that are present in all mammals. They are included in visual and somatosensory areas, and the auditory, olfactory and taste areas of the brain, as well as association areas of the cortex that integrate such information. Such sensory and association cortical regions eventually link up to the motor output neurons of the cortex, which in turn activate the muscle effector organs. What has made humans so unique is the massive expansion of prefrontal cortex, that region of the brain which integrates information across sensory dimensions and rationalises the behavioural consequences dependent on past experiences before activating the motor output neurons. The prefrontal cortex has substantial input to the ventral striatal brain reward centre in humans, making it possible for associations across behavioural events to themselves become fulfilling and rewarding.

WHAT IS SPECIAL ABOUT EPIGENETICS, AND WHY HAS THIS BEEN SO IMPORTANT IN THE EVOLUTIONARY ASCENT OF THE MATRILINE?

In addition to the genes which carry heritable genomic information across generations, there is another equally important 'epigenome' which determines how gene expression varies according to the numerous environmental perturbations that each individual's brain experiences. Such variations in gene expression result from the large number of ways that are available for 'activating' and 'silencing' of genes, mechanisms which go beyond the well-established textbook description of the transcription factors. These epigenetic mechanisms include the chemical modification (methylation, deacetylation, demethylation, etc.) of histone proteins that provide the structural support for, and access to DNA, as well as the direct silencing of DNA itself. The production of anti-sense non-coding RNAs, and the steroid hormones from the developing placenta, the adrenal glands and the gonads are all part of the epigenetic machinery that regulates the expression and silencing of genes. In the complex network of nuclear steroid hormone receptors, the long non-coding RNAs are emerging as critical determinants of hormone action (Klinge, 2015). These steroid RNA complexes associate with chromatin in a steroid hormone-dependent manner, an example of which is the Lnc-RNA 'Hotair' which shows increased recruitment to chromatin by oestrogen. Much of this work has focussed on breast and prostate cancer, but the action of many of the changes to non-coding RNAs are now being studied as part of normal physiology. These epigenetic factors form a group of important modifiers for gene expression that are classified as belonging to the non-heritable 'epigenome', and determine how one individual's expression of genes can vary from that of other individuals, even for those of monozygotic twins with an identical set of genes. Genetically identical twins that are reared apart still retain the same physical characteristics, but may differ in their lifestyle predispositions, the careers which they follow and in their psychiatric well-being. This outcome is special to one organ in particular, their brain, which evolution has designed to be epigenetically modifiable. Indeed, the brain's development transcends genetic determinism, and its functioning personifies the epigenetic processes underpinning learning and memory. During development and throughout life, the brain continues to make new interconnections and to lose others, a function that is dependent on the synaptic activity of neurons, activity which is environmentally driven. Not only does the brain adapt according to the environment in which it finds itself, but it can select its own preferred social environment and even make changes to its physical environment, thereby modifying its own developmental trajectory. Moreover, and especially in humans, the brain can carry these epigenetic changes through to the next and subsequent generations, not by epigenetic heredity, but through social communication, learning and memory. The rules and laws of societies can, to some extent, restrict and direct the developmental trajectories of the brain, while education usually broadens them. Once societies had recognised the importance for learning and the social interdependence of individuals, then humans gained the potential to direct both their own brain's destiny as well as that of others. The social collective of brains became greater than the sum of its individual brains.

Unique to both the neurons of the brain and to the developing cells of the germline is the capacity for reprogramming. The brain has evolved its own genetic variant for the germline reprogramming *Tet* genes, which code for the demethylating enzymes and which are best known for resetting the epigenome of each new generation (Hackett & Surani, 2013). In this way the brain is able to eliminate redundant 'short-term' memories, freeing up the neural capacity for forming new and relevant 'long-term' memories. This 'forgetting' may be considered a process which can sometimes be as important as the memory itself; a 'use it or lose it' strategy underpinning long-term memory. Children during the early developmental stages of their brain seem to remember everything, even that which is deemed inconsequential trivia by adults. In contrast, adults, via their prefrontal cortex, become more selective in what they actively submit to long-term memory.

The enlarged executive brain has played the latest and most significant role in shaping the evolutionary trajectory of humans, ameliorating them from the physical environment and expanding their focus to the social environment. Human brain function thus epitomises epigenetic regulation of its genes by transgenerational actions on behaviour and through the deployment of learning, memory and language. Critical periods in human brain development are also integral to infant attachment with the mother, thereby forming a secure base for developing other relationships and hence the future of their social well-being.

Any sex differences in the development of the human executive brain are far removed from the genetics of the Y-chromosome. As a consequence of the vast expansion of the human executive brain, human sexual proclivities may adopt a wide range of differing directions. These are often related to those experiences which have proved rewarding at an earlier period in life's sexual explorations by that individual, even if these activities are only a part of a self-gratifying imagination. In humans, sexuality is a function of a mind that has become uncoupled from reproduction. Indeed, most of human sexual behaviour is itself non-reproductive. This is not quite so true for maternal caring. The provisioning of nurture is closely linked to the placental hormones during pregnancy, and to infant suckling in the post-partum period. Indeed, suckling triggers the release of the hormone oxytocin, which not only acts on the mammary gland to release milk, but is released as a neuropeptide in the brain to promote and maintain maternal caring and mother–infant bonding. The mothering process may continue for as long as the infant requires, and is sustained through the continuous synthesis and release of oxytocin that is stimulated by each bout of the infant's suckling. However, the adult male body and mind has never experienced the presence of a placenta, its hormones, or the birth process and suckling, and although the male genome has helped to construct the foetal placenta, this genetic construction is primarily under the guidance of the female's imprinting of the genome.

The incredible investment that females make in producing children through sustaining a long pregnancy and post-partum care is successful due to a very special bonded relationship that develops between mother and child. This bonded relationship is activated by the birth process itself, an event which releases a flood of endorphins into the mother's brain. These endorphins are not only important for blocking the after pain of parturition, but are integral to activating that intense feeling of serenity after the birth of the baby. It is said that purity of love is never more intense than that of a mother for her newborn baby. This is not surprising, as the endogenous opioid endorphins play a very important role in brain reward, similar to the 'feel-good factor' that is said to occur when taking drugs of addiction. These drugs of addiction act on the same brain receptors that underpin social reward, and hence their abuse can be both addictive and extremely dangerous in usurping social relationships.

Having evolved such a large executive brain has enabled human behaviour to obtain a degree of emancipation from the hormonal language of their body's biology, but this may also have a downside. We humans mainly eat to avoid hunger, but we also succumb to the sensuous psychological rewards of food, thereby ignoring the hormonal satiety signals (leptin) from our body's fat deposits, and hence becoming predisposed to developing obesity and diabetes. The evolutionary progression of executive brain development has also empowered the human neocortex with a leading role in our reproductive success. This too may yet prove to be an evolutionary step too far. Having cautiously evolved successful brain development through natural selection at the genetic level, the very nature of human brain functioning has enabled the genetic basis of the brain's early creation to be superseded by the epigenetic regulation of its functioning. In humans it is the brain which becomes epigenetically tailored to match its social environment, even in the context of making important biological decisions. The problem arises as to how we can be sure these are the best or even the most appropriate decisions. The human brain is very much a servant of its social environment, with the most important socially secure attachment relationships developing being that between child and mother, leading to mother–infant bonding.

Human well-being starts with mother–infant bonding, which leads to the long-term security of infant attachment with the mother. It is from this secure base that young children expand their future social world. Worryingly, these social relationships are now being eroded by the use of electronic media. The iPhones, Twitter and Facebook of the modern world are replacing the physical interactions of social engagement, and enslaving the brains of our children into an electronic fantasy world, often very far removed from social reality. I am not convinced that fantasy rewards should replace the rewards of social reality, and especially the complex world of interpersonal social relationships. How ironic that the electronic devices which we have designed to improve communication have usurped the wealth of meaning that is communicated by body language, and the warmth of feeling this may generate through bodily contact.

REFERENCES

Ball, M., Carmody, M., Wynne, F., *et al.* (2009). Expression of pleiotrophin and its receptors in human placenta suggests roles in trophoblast life cycle and angiogenesis. *Placenta* 30: 649–53.

Berletch, J. B., Ma, W., Yang, F., *et al.* (2015). Escape from X inactivation varies in mouse tissue. *PLoS Genet.* 11: e1005079.

Broad, K. D., Curley, J. P. & Keverne, E. B. (2006). Mother–infant bonding and the evolution of mammalian social relationships. *Phil. Trans. R. Soc. B* 361: 2199–214.

Buckberry, S., Bianco-Miotto, T. & Roberts, C. T. (2014). Imprinted and X-linked non-coding RNAs as potential regulators of human placental function. *Epigenetics* 9: 81–89.

Cheng, T. L., Wang, Z., Liao, Q., *et al.* (2014). MeCP2 suppresses nuclear microRNA processing and dendritic growth by regulating *DGCR8/Drosha* complex. *Dev. Cell* 28: 547–60.

Colucci, F., Boulenouar, S., Keikbusch, J., *et al.* (2011). How does variability of immune system genes affect placentation? *Placenta* 32: 539–45.

Cortez, D., Marin, R., Toledo-Flores, D., *et al.* (2014). Origins and functional evolution of Y chromosomes across mammals. *Nature* 508: 488–93.
Curley, J. P. & Keverne, E. B. (2005). Genes, brains and mammalian social bonds. *Trends Ecol. Evol.* 20: 561–67.
Curley, J. P., Barton, S. C., Surani, A. M., *et al.* (2004). Co-adaptation in mother and infant regulated by a paternally expressed imprinted gene. *Proc. R. Soc. Lond. B Biol. Sci.* 271: 1303–09.
Dunbar, R. (2003). Psychology. Evolution of the social brain. *Science* 302: 1160–61.
Ferguson-Smith, A. C. (2011). Genomic imprinting: the mergence of an epigenetic paradigm. *Nat. Rev. Genet.* 12: 565–75.
Gendrel, A-V. & Heard, E. (2014). Noncoding RNAs and epigenetic mechanisms during X-chromosome inactivation. *Annu. Rev. Cell Dev. Biol.* 30: 4.1–4.20.
Goldman-Rakic, P. S. (1987). Development of cortical circuitry and cognitive function. *Child Dev.* 58: 601–22.
Gribnau, J. & Grootegoed, J. A. (2012). Origin and evolution of X chromosome inactivation. *Curr. Opin. Cell Biol.* 24: 397–404.
Hackett, J. A., Sengupta, R., Zylic, J. J., *et al.* (2013). Germline DNA demethylation dynamics and imprint erasure through 5-hydroxymethylcytosine. *Science* 339: 448–52.
Hackett, J. A. & Surani, M. A. (2013). Beyond DNA: programming and inheritance of parental methylomes. *Cell* 153: 737–39.
Hajkova, P., Jeffries, S. J., Lee, C., *et al.* (2010). Genome-wide reprogramming in the mouse germ line entails the base excision repair pathway. *Science* 329: 78–82.
Hatanake, Y., Shimizu, N., Nishikawa, S., *et al.* (2013). GSE is a maternal factor involved in active DNA demethylation in zygotes. *PLos ONE* 8: e60205.
Hutter, B., Bieg, M., Helms, V., *et al.* (2010). Imprinted genes show unique patterns of sequence conservation. *BMC Genomics* 11: 649.
Jedrusik, A., Cox, A., Wicher, K. B., *et al.* (2015). Maternal-zygotic knockout reveals a critical role of *Cdx2* in the morula to blastocyst transition. *Dev. Biol.* 398: 147–52.
Karre, K. (2008). Natural killer cell recognition of missing self. *Nat. Immunol.* 9: 477–80.
Keverne, E. B. (2004). Understanding well-being in the evolutionary context of brain development *Phil. Trans. R. Soc. Lond. B* 359: 1349–58.
  (2006). Trophoblast regulation of maternal endocrine function and behaviour. In: A. Moffett, C. Loke & A. McLaren (eds.), *Biology and Pathology of Trophoblast*. New York, NY: Cambridge University Press, pp. 148–63.
  (2013). Importance of genomic imprinting in the evolution and development of the maternal brain In: D. Pfaff & Y. Christen (eds.), *Multiple Origins of*

Sex Differences in Brain. Neuorendocrine Functions and their Pathologies. Berlin: Springer-Verlag, pp. 21–34.

(2014). Mammalian viviparity: a complex niche in the evolution of genomic imprinting. *Heredity* 113: 138–44.

(2015). Genomic imprinting, action, and interaction of maternal and fetal genomes. *Proc. Natl Acad. Sci. USA* 112: 6834–40.

Klinge, C. M. (2015). Estrogen action: receptors, transcripts, cell signaling, and non-coding RNAs in normal physiology and disease. *Mol. Cell. Endocrinol.* 418: 191–92.

Kolb, B., Mychasiuk, R., Muhammad, A., et al. (2012). Experience and the developing prefrontal cortex. *Proc. Natl Acad. Sci. USA* 109: 17186–93.

Leseva, M., Knowles, B. B., Messerschmidt, D. M., et al. (2015). Erase–maintain–establish: natural reprogramming of the mammalian epigenome. *Cold Spring Harb. Symp. Quant. Biol.* 60: 155–63.

Li, X., Ito, M., Zhou, F., et al. (2008). A maternal–zygotic effect gene, Zfp57, maintains both maternal and paternal imprints. *Dev. Cell* 15: 547–57.

Loke, Y. W. (2013). *Life's Vital Link: The Astonishing Role of the Placenta.* Oxford: Oxford University Press.

& King, A. (2000). Immunological aspects of human implantation. *J. Respod. Fertil. Suppl.* 55: 83–90.

Marchetti, F., Essers, J., Kanaar, R., et al. (2007). Disruption of maternal DNA repair increases sperm-derived chromosomal aberrations. *Proc. Natl Acad. Sci. USA* 104: 17725–29.

Meng, L., Ward, A. J., Chun, S., et al. (2015). Towards a therapy for Angelman syndrome by targeting a long non-coding RNA. *Nature* 518: 409–12.

Merzouk, S., Deuve, J. L., Dubois, A., et al. (2014). Lineage-specific regulation of imprinted X inactivation in extraembryonic endoderm stem cells. *Epigenetics Chromatin* 7: 11.

Muir, A., Lever, A. M. & Moffett, A. (2006). Human endogenous retrovirus-W envelope (syncytin) is expressed in both villous and extravillous trophoblast populations. *J. Gen. Virol.* 87: 2067–71.

Murphy, P. J. & Cairns, B. R. (2016). Genome-wide DNA methylation profiling in zebrafish. *Methods Cell. Biol.* 135: 345–59.

Potok, M. E., Nix, D. A., Parnell, T. J., et al. (2013). Reprogramming the maternal zebrafish genome after fertilization to match the paternal methylation pattern. *Cell* 153: 759–72.

Prudhomme, J. & Morey, C. (2016). Epigenesis and plasticity of mouse trophoblast stem cells. *Cell Mol. Life Sci.* 73: 757–74.

Renfree, M. B., Suzuki, S. & Kaneko-Ishino, T. (2013). The origin and evolution of genomic imprinting and viviparity in mammals. *Phil. Trans. R. Soc. B* 368: 20120151.

Sandhu, K. S. (2010). Systems properties of proteins encoded by imprinted genes. *Epigenetics* 5: 627–36.

Santos, F., Peat, J., Burgess, H., et al. (2013). Active demethylation in mouse zygotes involves cytosine deamination and base excision repair. *Epigenetics Chromatin* 6: 39.

Sato, Y., Shinka, Y., Sakamoto, K., et al. (2010). The male-determining gene *SRY* is a hybrid of *DGCR8* and *SOX3*, and is regulated by the transcription of CP2. *Mol. Cell. Biochem.* 337: 267–75.

Schultz, W. (2016). Reward functions of the basal ganglia. *J. Neural Transm. (Vienna)* 123: 679–93.

Sun, S., Payer, B., Namekawa, S. H., et al. (2015). Xist imprinting is promoted by the hemizygous (unpaired) state in the male germ line. *Proc. Natl Acad. Sci. USA* 112: 14415–22.

Tang, W. W., Kobayashi, T., Irie, N., et al. (2016). Specification and epigenetic programming of the human germ line. *Nat. Rev. Genet.* 17: 585–600.

Varrault, A., Gueydan, C., Delalbre, A., et al. (2006). Zac1 regulates an imprinted gene network critically involved in the control of embryonic growth. *Dev. Cell* 11: 711–722.

Wu, J., Huang, B., Yin, Q., et al. (2016). The landscape of accessible chromatin in mammalian preimplantation embryos. *Nature* 534: 652–57.

Yang, P., Wu, W. & Macfarlan, T. S. (2015). Maternal histone variants and their chaperones promote paternal genome activation and boost somatic cell reprogramming. *Bio Essays* 37: 52–59.

Index

addiction
    effect on mother–infant bonding, 132
addictive behaviour, 98, 101, 103, 201
adrenogenital syndrome (AGS), 11–12
Adult Assessment Interview, 130
ageing
    epigenetic effects, 33–35
    influence of sirtuins, 34–35
    role of sirtuins, 33
Agouti gene, 27–28, 62
Agouti mouse, 27–28, 62
alloparenting, 119–20, 187
allopregnenalone, 38
Alzheimer's disease, 24
Amami spiny rat
    lack of a Y-chromosome, 2
amygdala, 37, 98, 105, 118
Ancient Greeks
    views on differences between the sexes, vii
androgen receptor gene
    location on the X-chromosome, 6–7
androgenetic chimera embryo studies, 55–60
androgenetic embryo studies, 55
Angelman's syndrome, 74, 152, 187
    genomic imprinting effects, 58
Aristotle, vii
attachment
    assessment in adults, 130–31
    assessment of children, 130
    behaviour in children, 128–33
    effect of maternal depression, 131–32
    effect of maternal drug addiction, 132
    inter-generational transmission, 131
    lifelong influence of early attachment, 131
    mechanisms, 132–33
autism spectrum disorders (ASD), 9, 168

BDNF gene, 41
behavioural disorders
    risk during puberty, 103–08

bipolar disorder
    creativity and, 45–46
    during puberty, 104–05
    gender differences in, 135
blastocyst, 192
Bowlby, John, 128–29
brain development
    chimeric embryo studies, 55–60
    developmental disorders, 152
    differences in male and female brains, 37–40
    disorders related to genomic imprinting, 80
    during puberty, 87–89
    effects of social adversity, 43–45, 126–28
    endorphin receptors in the brain, 101–02
    epigenetic influences, 35–40
    epigenetic modifications, 198–200
    expression of retrotransposons, 31–32
    extended maternal care, xi–xii
    extended maturation period in humans, 90–91
    extended period of post-natal growth, xi–xii
    genes which escape reprogramming, 31–32, 79–80, 182–83
    genomic imprinting studies, 55–60
    grey matter and white matter, 97
    hormonal modifiers of the female brain, 37–40
    hypothalamus, 189
    in utero development, 138–39
    influence of the social environment, 40–41, 43–45
    influences on neocortical development, 40–43
    Klinefelter's syndrome (XXY), 9
    learning and memory, 46
    limbic brain, 36
    masculinisation of the male brain by testosterone, 139–40

brain development (cont.)
  mechanisms underpinning learning and memory, 41–42
  myelination of neurons, 96
  neocortical asymmetry, 97–98
  neocortical development, 140–41
  neocortical reorganisation in late adolescence, 35–36
  olfactory system, 42–43
  post-natal maternal care, 139
  reorganisational changes during puberty, 95–102
  retrotransposons and, 28
  role of epigenetics, 193–97
  role of gene regulatory mechanisms, 47–48
  role of the X-chromosome, 8–9
  sex differences in reorganisation at puberty, 96–102
  stages of, 140–41
  synaptic plasticity, 41–42
  variability between individual brains, 48
brain evolution
  ability to override hormonal signals, xi–xii
  brain's ability to shape its own evolution, xi–xii
  co-adaptation with placental evolution, 64–65, 74–75, 145–48, 188–93
  consequences of, 201–02
  development of visual information processing, 114–17
  emotional brain, 151–52, 193–94
  executive brain, 36–37, 149–52
  expansion of the neocortex, 194–97
  forebrain, 58–60
  functions of the neocortex, 36–37
  influence of complex social relationships, 117–26
  influence of placental viviparity development, 145–47
  limbic brain, 193–94
  maternal emotional brain co-adaptation with the placenta, 152–59
  matrilineal influence on the neocortex, 162–70
  role of epigenetics, 193–97
  sex differences and, x–xii
  shift from hormonal to social regulation of behaviour, 36–37
  social relationships and, xi–xii
brain function
  frontal neocortex, 151
  influence of placental hormones on the female brain, 37–40
  neural reward system, 151
  opioid receptors in the brain, 101–02
  theory of mind, 151
  transmission of information to other brains, 46–47
  ventral striatum, 151
brain reward system
  endorphins and social cohesion, 120–23
Brenner, Sydney, 20
Broca, Paul, ix

cancer
  retrotransposons and, 27
  role of Lnc-RNAs, 198
Cbx2 gene, 2, 4, 23, 174–75
Cdx2 gene, 79, 184, 189
chemosensory system, 68, 112
  decline in importance in large brained-mammals, 113–14
chimeric embryo studies, 55–60
chimpanzees, 165–66
chromatin polycomb complexes, 23
chromosomes, 1
  sex chromosomes, 1–2
co-adaptive gene expression, 67–68
colour vision, 114
conduct disorders in adolescent males, 99
corpus callosum, 99
corticosteroids, 90
creativity
  and mood disorders, 45–46
Crick, Francis, 19
cultural development, 47

Darwin, Charles, ix
depression
  creativity and, 45–46
  effect on mother-infant bonding, 131–32
  gender differences in brain activity, 135
  related to puberty, 103–04
DGCR8 gene, 14, 159–60, 175
diffusion weighted imaging (DFI), 105
dihydrotestosterone, 12
Dlk1 gene, 61
Dlk-Dio3 cluster, 64

DNA
  histone protein framework modifications, 21–22
  'memory' marks (methylation and demethylation), 21–22
  methylation profiles in the fertilised egg, 27
  non-coding functional regulatory elements, 21–22
  regulation of gene expression, 21–22
  tight packing of sperm DNA, 73
DNA CpGs, 21
DNA methylation
  silencing of genes, 21
DNA methyltransferases, 41–42
DNA mismatch repair, 70, 72–73, 78
  maternal control in mammals, 183–84
DNA repeat sequences, 20
DNA structure
  genetic code, 19
  human genome sequence, 20–21
  non-coding DNA, 20–21
Dolly the sheep, 29
dopamine, 125
Drosha/Dgcr8 complex, 53, 191
drugs of abuse, 101
duck-billed platypus
  sex determination by many genes, 2
Dutch famine of 1944
  low birth weight in second generation children, 29

eating disorders, 103
Edwards, Bob, 54
Eif2s3y gene, 3
emotional awareness
  developments a puberty, 98–99
emotional brain, 36
  evolution, 151–52, 193–94
emotional brain (maternal)
  co-adaptation with the foetal placenta, 152–59
emotional cognition
  brain development and, 100–01
endorphins, 125
  receptors in the brain, 101–02
  role in maternal caring, 200–01
  role in promoting social cohesion, 120–23
  β-endorphin, 163

epigenetic clock
  ageing and, 33–35
  role of sirtuins in ageing, 33
epigenetic drift, 34
epigenetic reprogramming
  controlling role of the maternal genome, 32–33
  genes which escape reprogramming, 31–32
  germline cells, 28–33
  oocytes, 29
  paternal genome reprogramming stages, 29–31
  stem cells, 29
epigenetic sex-determining factors, 2
epigenetic variability, 34
epigenetics
  DNA 'memory' marks (methylation and demethylation), 21–22
  effects of adverse environments, 25–26
  germline reprogramming, 25–27
  histone code, 21–22
  influence on brain development and function, 35–40
  influence on matriline evolution, 198–202
  influence on neocortical development, 40–43
  lifetime effects on individuals, 25–26
  mechanisms, 198–200
  multiple levels of gene regulation, 25
  regulation of gene expression, 21–22
  regulatory controls on non-coding RNA, 24–25
  role in brain development, 193–200
  role in brain evolution, 193–97
  transgenerational effects, 25–26
  transgenerational inheritance, 182
epigenome, 198
  male and female differences in germline cells, 29–31
evolution of genomic imprinting
  theories of, 65–69
executive brain, 36–37
  evolution, 149–52
extended phenotype theory, 69
Ezh2 gene, 23

female reproductive strategy, 67–69
fertilisation
  replacement of sperm protamines with histones, 30

fertilised egg
  DNA methylation profiles, 27
fish
  egg-laying reproductive strategy, 179
  germline reprogramming, 179–80
foetal genome
  co-adaptation with the maternal genome, 37–40
foetal hormones
  influence on the mother's physiology and behaviour, 37–40
foetal programming
  induced by maternal stress, 154
forebrain evolution
  effects of genomic imprinting, 58–60
FoxL2 gene, 4
Franklin, Rosalind, 19
Freud, Sigmund, viii
FTO gene (fat mass obesity gene), 24–25
functional magnetic resonance imaging (fMRI), 105

GABA receptors, 41
gene expression
  epigenetic regulation, 21–22
  multiple levels of epigenetic regulation, 25
  role of non-coding RNA, 20
gene expression regulation
  role of non-coding DNA, 20–21
gene repression
  role in sex determination, 3–4
genes
  silencing by DNA methylation, 21
genetic code
  structure of DNA, 19
genetic hub, 25
genomic imprinting, 31
  androgenetic chimera embryo studies, 55–60
  androgenetic embryo studies, 55
  Angelman's syndrome, 58
  brain and placenta co-adaptation, 64–65
  brain disorders associated with, 80
  chimeric embryo studies, 55–60
  co-adaptation in maternally imprinted genes, 64–65
  co-adaptive evolution of structures, 74–75
  co-adaptive gene expression, 67–68
  definition, 52
  disorders caused by loss of imprinting, 73–74
  DNA methylation marking, 52–53
  DNA mismatch repair, 70, 72–73
  evolutionary origins, 60–65
  evolutionary perspective, 184–88
  evolutionary theories, 65–69
  historical perspective, 53–60
  horse x donkey hybrids, 53
  hybrid mammalian crosses, 53–54
  imprint control region (ICR), 52–53, 66–67, 80
  imprinted gene networks, 73–75
  leading role of the matriline, 67–69
  lion x tiger hybrids, 53–54
  male germline, 62–63
  marsupials, 60, 63
  matrilineal stabilising of imprinted genes, 70–73
  mechanisms, 52–53
  mechanisms for evolutionary success, 80–82
  niche construction theory, 69
  number of genes regulated in humans, 63
  parent of origin expression of autosomal genes, 52
  parental conflict theory, 66
  parthenogenetic chimera embryo studies, 55–60
  parthenogenetic embryo studies, 54–55
  Prader–Willi syndrome, 58
  purifying selection of mutations, 70
  reprogramming of the imprinted genome, 76–80
  role in forebrain evolution, 58–60
  role of retrotransposons, 60–63
  stability of imprinted genes, 70–73
genomic reprogramming
  genes which escape reprogramming, 79–80
  reprogramming of the imprinted genome, 76–80
germline cells
  epigenetic reprogramming, 28–33
  male and female epigenetic differences, 29–31
germline reprogramming, 25–27, 178–83
  brain development and, 182–83
  DNA methylation profiles in the fertilised egg, 27
  early genetic landmarks, 179

evolutionary perspective, 179–82
genes which escape
    demethylation, 182–83
maternal lead in mammals, 180–82
paternal lead in fish, 179–80
purpose of, 178
zebrafish, 180
glucose homeostasis, 33
gonadotrophin-releasing hormone
    (GnRH), 92, 94
gonads
    development influenced by sex
        chromosomes, 1–2
    production of sex hormones, 1–2
Gse gene, 184
guevodoce children, 12–13
Gurdon, John, 29, 76

H19 Lnc-RNA, 64
HERV gene, 192
heterogametic sex, 1–2
hippocampus, 41–42, 117
histone code, 21–22
histone deacetylases
    sirtuins, 33
HLA epitopes, 192
homeothermy, 111
homogametic sex, 1
hormonal signals
    ability of the human brain to override, xi–xii
horse x donkey hybrids
    genomic imprinting effects, 53
Hotair Lnc-RNA, 23–24, 198
human genome
    non-coding DNA, 20–21
    number of genes in, 20
    sequencing, 20–21
human hybrids, 54
human sexuality, 200
human social evolution, 47
    history of, 133–35
human societies
    significance of the matriline, 136–38
Huxley, Thomas, viii–ix
hybrid mammalian crosses, 53–54
hypothalamus, 37, 118, 189
    functions during puberty, 89–90
    maternal, 190–91
hypothamic–piuitary–gonadal axis
    (HPG-axis), 92

Igf2 gene, 64, 66–67
Igf2r gene, 66
immune system
    foetal placenta avoidance of
        rejection, 191–93
imprint control region (ICR), 52–53,
    66–67, 80
in utero development, xi–xii, 14, 33, 111
    brain development, 138–39
    evolutionary role of genomic
        imprinting, 64–65
    evolutionary selection pressures, 39
    heritable features associated with, 39
in vitro fertilisation technique, 54
infanticide of females, 136–38
insulin, 24
inter-generational co-adaptation
    mother's brain and foetal placenta, 188–93
    three matrilineal genomes, 145–48
intergenerational genomic
    co-adaptation, xii
intergenerational inheritance, 43–48
intergenerational transmission of
    attachment, 131

Jacob, François, 19
Jost, Alfred, viii

kisspeptin
    role in puberty onset, 91–93
Klinefelter's syndrome (XXY), 9

L1 transposons, 27–28
lactation, 111
language acquisition, 99
lateral geniculate nucleus, 116
leadership, 47, 134–35
    patrilineal societies, 136–38
learning and memory, 46
    underpinning mechanisms, 41–42
leptin, 88–89, 92, 94
    role in puberty onset, 94
Leydig cells, 39, 160, 175
lifespan
    effects of calorie restriction, 35
    influence of sirtuins, 34–35
limbic brain, 57, 118
    evolution, 193–94
    interactions with the neocortex, 36
    structure and functions, 36

lion x tiger hybrids
  genomic imprinting effects, 53–54
long interspersed nuclear elements (LINES), 60
long non-coding RNA (Lnc-RNA), 23–24, 176, 198
LTR (long terminal repeat) retrotransposons, retroviral elements, 60–61
Lwoff, André, 19

M6A demethylase enzyme, 24
male brain
  masculinisation, 157–62
  masculinising effect of testosterone, 139–40
male germline
  genomic imprinting, 62–63
male reproductive strategy, 67–69
male-typical behaviours
  hormonal influences, 113
males
  barriers to developing maternal behaviour, 157–59
mammalian evolution
  role of the matriline, 172–78
mammals
  evolution of sex differences, 14–16
manic depression
  creativity and, 45–46
marsupials, 74, 160, 185
  genomic imprinting, 60, 63
  placentation, 60
masculinisation of the male brain, 139–40, 157–59
  evolutionary perspective, 159–62
masculinity
  dependence on the female germline, 7–8
  role of the X-chromosome, 4–5
  role of the Y-chromosome, 5
maternal care
  endorphins and, 200–01
  evolutionary perspective, 111–17
  extended postnatal care, xi–xii
maternal genome
  co-adaptation with the foetal genome, 37–40
  control of epigenetic reprogramming, 32–33
  control of paternal genome reprogramming, 30–31

maternal stress-induced foetal programming, 154
matriline
  evolutionary influence of epigenetics, 198–202
  influence on mammalian evolution, xii
  influence on neocortical evolution, 162–70
  leading role in genomic imprinting, 67–69
  role in determining masculinity, 7–8
  stabilising imprinted genes, 70–73
matrilineal genome
  contribution to development of sex differences, xii
  influence on sex differences, ix–x
  intergenerational co-adaptive evolution, 145–48
matrilineal societies, 136–38
Mead, Margaret, 86
MeCP2 enzyme, 168
MeCp2 gene, 9–10
MeCp2 protein mutations, 28
medio-basal hypothalamic forebrain, 57
messenger RNA, 19, 23
metabolic disorders, 182–83
metabolism
  regulatory role of sirtuins, 33
micro-RNA, 23, 62–63, 191
migration
  gender differences in response to, 135
mitochondria
  fate of sperm mitochondria, 31, 73
  maternal line inheritance, 31, 73
MKRN3 gene, 92–93
Monod, Jacques, 19
monogamy in small-brained mammals, 122–24
monotremes, 60–61, 160, 185
mood disorders
  creativity and, 45–46
motherhood
  influence on mammalian reproductive success, xii
mother–infant bonding, xi
  adult attachment assessment, 130–31
  alloparenting, 119–20
  attachment behaviour in children, 128–33
  brain evolution and, xii

child attachment types, 130
  effect of maternal depression, 131–32
  effect of maternal drug addiction, 132
  effect of separation of mother and
    infant, 126–28
  evolutionary perspective, 111–17
  formation and regulation of primate social
    bonds, 117–26
  foundations for future interpersonal
    relationships, 128–33
  influence on brain development, 139
  intergenerational transmission of
    attachment, 131
  lifelong influence of early attachment, 131
  mechanisms of attachment, 132–33
  role of oxytocin, 112–13
  Strange Situation child assessment, 130
mother's brain
  co-adaptation with the foetal
    placenta, 188–93
motivational behaviour
  evolution of, 59–60

neocortex, 57, 117
  evolutionary expansion, 194–97
  evolutionary influence of the
    matriline, 162–70
  functions, 36–37
  interactions with other parts of the
    brain, 36–37
  late adolescent reorganisation
    phase, 35–36
neocortical asymmetry, 97–98
neocortical development, 140–41
  epigenetic influences, 40–43
neotany, 167
Netsilik Eskimos, 137
neural cell mosaicism, 28
niche construction theory of genomic
  imprinting, 69
NK (natural killer) cells (maternal), 192–93
NMDA receptors, 41
non-coding DNA, 20–21
  role in gene expression regulation, 20–21
non-coding RNA
  epigenetic regulatory controls on, 24–25
  in imprinted gene clusters, 53
  long non-coding RNA
    (Lnc-RNA), 23–24
  role in regulation of gene expression, 20

obesity, 182
obsessive compulsive behaviours, 101,
  103, 127
oestrogen, 90, 97, 113, 198
olfactory system, 41–42, 112, 116
  decline in importance in large-brained
    mammals, 113–14
  neural plasticity, 42–43
oocytes
  epigenetic reprogramming, 29
opioid receptors in the brain, 101–02
ovaries
  development of, 3–4
oxytocin, 38, 90, 120, 123, 163, 200
  role in mother–infant bonding, 112–13

p53 gene, 34
Papua New Guinea, 136
parental conflict theory of genomic
  imprinting, 66
parthenogenetic chimera embryo
  studies, 55–60
parthenogenetic embryo studies, 54–55
paternal DNA
  mismatch repair after fertilisation, 30
paternal genome
  reprogramming controlled by the
    maternal genome, 30–31
  stages of epigenetic reprogramming, 29–31
patrilineal societies, 136–38
Peg3 gene, 74, 187
Peg10 gene, 61
Peg11 gene, 61
Pgc7 gene, 184
pheromones, 112
  decline in importance in large-brained
    mammals, 113–14
pituitary gland
  functions during puberty, 89–90, 92
placenta, 77, 79
  avoidance of rejection by the maternal
    immune system, 191–93
  co-adaptation with the mother's brain,
    152–59, 188–93
  co-adaptive evolution with the brain,
    74–75, 145–48
  development dependent on maternal
    alleles, 33
  development of placentation, 111
  effects of the imprinted genome, 186

placenta (cont.)
  evolution of, 145–47
  foetal genetic origin, 188–93
  influence on mammalian evolution, xii
  marsupials, 60
placental autophagy, 157
placental development
  genomic imprinting, 53
  role of the X-chromosome, 8–9
placental dysfunctions
  genomic imprinting, 53
placental evolution
  co-adaptation with brain evolution, 64–65
  syncitin transposon genes, 60
placental genome
  transgenerational co-adaptive
    effects, 37–40
placental hormones, 113
  functions of, 186
  influences on the female brain, 37–40
placental mammals
  evolution of, 160
placental stem cells, 40
polycomb repressive complexes, 3, 20, 23
Prader–Willi syndrome, 56, 74, 93, 152
  genomic imprinting effects, 58
prairie vole
  monogamous pair-bonding, 122–24
pre-optic area, 118
progesterone, 90, 113
  placental, 37
protamine packing of sperm DNA, 30, 73, 181
  replacement with histones at fertilisation, 30
pseudogenes, 20
psychiatric disorders
  risk during puberty, 103–08
psychopathology
  creativity and, 45–46
puberty
  bipolar disorder and, 104–05
  body fat influence on onset, 94
  brain development during, 87–89
  brain maturation and, 94–95
  consequences of early onset, 93–95
  cultural aspects, 86–87
  delayed onset, 92
  depression related to, 103–04
  different male and female sensitivity to kisspeptin, 92–93
  effects of stress, 90
  evolutionary perspective, 91
  extended period for brain maturation, 90–91
  functions of the pituitary gland, 89–90, 92
  genetic mutations affecting onset, 93
  hormonal changes, 89–90
  Kiss 1 and Kiss 2 neurones, 92–93
  manifestation of schizophrenia, 104
  nutrition and, 87–89
  precocious, 92–93
  progressively earlier onset, 87–89
  reorganisational changes in the brain, 95–102
  risk of behavioural disorders, 103–08
  risk of psychiatric disorders, 103–08
  role of kisspeptin in puberty onset, 91–93
  role of leptin, 94
  role of the hypothalamus, 89–90
  sex differences in brain reorganisation, 96–102
  social relationships and, 101–02
  suicide risk, 103–04
  turbulence and stress of, 86–87
purifying selection of genetic mutations, 70

rapamycin signalling pathway, 35
Reelin gene, 41
reproductive strategies
  differences between males and females, 67–69
reptiles
  epigenetic sex-determining factors, 2
retrotransposons
  Agouti gene, 27–28
  brain development and, 28
  definition, 27
  evolutionary role, 60–63
  expression in the brain, 31–32
  role in brain evolution, 27–28
retroviral DNA
  genomic imprinting, 62–63
Rett syndrome, 9–10, 27–28, 61
reward system in the brain, 151
ribosomal RNA, 23
RNA
  forms of, 23
  long non-coding RNA (Lnc-RNA), 23–24
  messenger RNA, 19
  methylation and demethylation, 24–25
  micro-RNA, 23

structure, 23
Rtl1 gene, 61

salamander genome, 20
schizophrenia, 28, 182
  manifestation during puberty, 104
self-sacrifice for the benefit of others, 138
sex chromosomes
  influence on development of gonads, 1–2
  Klinefelter's syndrome (XXY), 9
  mechanisms for sex determination, 1–2
  Turner's syndrome (XO), 8–9
sex determination
  arrangements of sex chromosomes, 1–2
  differences across species, 2
  duck-billed platypus, 2
  early influence of sex steroid
    hormones, 10–14
  epigenetic factors, 2
  leading role for the matriline, 172–78
  role of gene repression, 3–4
  role of the *SRY* gene, 2–3
  species with no Y-chromosome, 2
sex differences
  biological status of the female, 172–78
  brain evolution and, x–xii
  brain reorganisation at puberty, 96–102
  challenges to the male dominance
    view, x–xii
  contribution of the matrilineal
    genome, xii
  evolution of, 14–16
  evolutionary origins, 172–78
  historical views on, vii–ix
  influence of the matrilineal genome, ix–x
  influential male-biased opinions on, vii–ix
  social view and biological reality, ix–x
  views in Ancient Greece, vii
  views of the Victorians, viii–ix
sex hormone receptor genes, 1
sex hormones
  changes at puberty, 89
  early influence on sex
    determination, 10–14
  production by gonads, 1–2
Shakespeare, William, 86
short interspersed nuclear elements
  (SINES), 60
Sirt6 gene, 35
sirtuins, 35
  links with cellular senescence, 34–35
  role in ageing, 33
  role in regulation of cell metabolism, 33
single nuclear polymorphisms (SNPs), 47
social adversity
  effects on brain development,
    43–45, 126–28
social bonding
  brain development and, 101–02
  formation and regulation of primate social
    bonds, 117–26
  monogamy in small-brained
    mammals, 122–24
social cohesion
  promoting by endorphins, 120–23
social environment
  influence on brain development,
    40–41, 43–45
social evolution
  history of, 133–35
social grooming
  endorphin release, 120–23
  promotion of social cohesion, 120–23
social learning, 117–20
social living
  brain evolution and, xi–xii
social relationships
  changes at puberty, 101–02
  foundations in mother–infant
    bonding, 128–33
  influence on brain evolution, 117–26
social status, 119–20, 140, 162
societies
  significance of the matriline, 136–38
Sox2 gene, 28
Sox3 gene, 14, 159, 175
Sox9 gene, 4, 175
sperm
  protamines replaced with histones at
    fertilisation, 30
  stages of epigenetic
    reprogramming, 29–31
sperm DNA, 78
  male zebrafish, 180–81
  mismatch repair, 78
  tight packing by protamines, 30, 73, 181
SRY gene, 3–4, 15
  evolution of, 159–60
  functions, 159–60, 175–76
  loss in the Amami spiny rat, 2
  role in sex determination, 2–3
Stella gene, 184

stem cells
  epigenetic reprogramming, 29
  placental, 40
stereotypical behaviours, 126
stress
  effects during puberty, 90
  maternal stress-induced foetal programming, 154
stress hormones, 90
striatum, 57, 117
Suchi-ishi retrotransposon, 61
suicide
  risk during puberty, 103–04
Surani, Azim, 54
synaptic plasticity, 41–42
syncitin transposon genes, 60
syncytins, 192
syncytiotrophoblast, 192

tammar wallaby, 60
test-tube baby technology, 54
testes
  development of, 3–4
testicular feminisation syndrome (Tfm), 6–7, 13
testosterone, 90, 97–98, 113, 160
  effects of, 13
  functions of, 6
  influence on sex differences, x
  masculinising effect on the male brain, 139–40
  X-chromosome location of the androgen receptor gene, 6–7
Tet1,2,3 methylcytosine dioxygenases, 42–43
theory of mind, 151
Thomas Aquinas, St, viii
Thucydides, vii
tiger x lion hybrids
  genomic imprinting effects, 53–54
transcranial direct current stimulation (tDCS), 105–06
transcription factors, 21
  for maternal-specific gene networks, 74
transfer RNA, 23
transgenerational epigenetic inheritance, 182
transposons, 20
  evolutionary role, 28, 60–63, *see also* retrotransposons

Trim28 gene, 27, 61
trophectoderm, 33, 77, 79, 184, 192
Turner's syndrome (XO), 8–9

UBE3a gene, 187

vasopressin, 90, 113
ventral striatal brain reward regions, 106
ventral striatum, 98
Victorian era
  views on differences between the sexes, viii–ix
visual information
  importance for large-brained mammals, 114–17
visual relay nucleus, 116
viviparity, xi–xii, 14, 33, 111
  evolution of, 145–47
  evolutionary role of genomic imprinting, 64–65

Waddingtonism, 108
Watson, James, 19
Wilkins, Maurice, 19
WNT4 gene, 15
working memory, 99

X-chromosome
  androgen receptor gene, 6–7
  evolutionary acquisition of genes, 4–5
  evolutionary feminisation, 6
  genes which are present on both X- and Y-chromosomes, 6
  homologous XY gene pairs, 7
  male inheritance from the mother, 4–5
  mechanisms for sex determination, 1–2
  role in determining masculinity, 4–5, 7–8
  specialisation for male reproduction, 4–5
X-inactivation, 8, 10, 176–78
  effects of skewed inactivation, 9
  Xist RNA, 20
X-linked intellectual disability, 9
Xist gene, 8
Xist RNA, 176–77
  role in X-inactivation, 20
XY homologous gene pairs, 7
XY sex reversal, 15

Yamanaka, Shinya, 29, 76
Y-chromosome
  absence of DNA repair enzymes, 3
  effects of having a single
    Y-chromosome, 2–3
  evolutionary demasculisation, 6
  evolutionary loss of genes, 3
  father-to-son inheritance, 5
  genes which are present on both X- and
    Y-chromosomes, 6
  higher rate of mutations, 3, 5
  influence on development of
    testes, 1–2
  influence on sex differences, x
  lack of DNA mismatch repair, 5–6
  maintenance of homologous XY gene
    pairs, 7
  mechanisms for sex determination, 1–2
  operation of natural selection on, 5–6
  potential to disappear in humans, 3
  role in determining masculinity, 5
  role of the *SRY* gene, 2–3
  species which lack a Y-chromosome, 2
  stablity of remaining genes, 3

zebrafish
  germline reprogramming, 180
Zfp57 gene, 27, 61, 184